T0358274

Ambrogio Leone's *De Nola*, Venice 1514

Brill's Studies in Intellectual History

General Editor

Han van Ruler (*Erasmus University Rotterdam*)

Founded by

Arjo Vanderjagt

VOLUME 284

The titles published in this series are listed at *brill.com/bsih*

Ambrogio Leone's *De Nola,* Venice 1514

Humanism and Antiquarian Culture in Renaissance Southern Italy

Edited by

Bianca de Divitiis
Fulvio Lenzo
Lorenzo Miletti

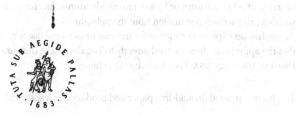

BRILL

LEIDEN | BOSTON

Cover illustration: Ambrogio Leone, Girolamo Mocetto, *Nola Vetus*. Detail.

The Library of Congress Cataloging-in-Publication Data is available online at http://catalog.loc.gov

Typeface for the Latin, Greek, and Cyrillic scripts: "Brill". See and download: brill.com/brill-typeface.

ISSN 0920-8607
ISBN 978-90-04-37577-2 (hardback)
ISBN 978-90-04-37578-9 (e-book)

Contents

Acknowledgements

This volume is the result of an interdisciplinary research carried out as part of the European Research Council research project *Historical Memory, Antiquarian Culture and Artistic Patronage: Social Identities in the Centres of southern Italy between the Medieval and Early Modern Period* (2011–2016), FP7/2007–2013, ERC Grant Agreement n° 263549.

For the volume the project was expanded with the addition of essays by Eugenio Imbriani and Giuliana Vitale, to whom the project team would like to extend its thanks.

A special thanks also goes to Stephen Parkin for contributing with an essay and for translating and revising the texts. We are grateful to all the HistAntArtSI research team, and in particular Antonio Milone and Francesca Buccheri for their generous assistance.

We would like to thank the Seminario Diocesano of Nola and the University of Venice Ca' Foscari for hosting two seminars on the *De Nola*. Francesco Aceto, Howard Burns, Francesco Caglioti, Donatella Calabi, Caroline Elam, Seth Fagen, Jill Kraye, Francesco Senatore, Antonia Solpietro and William Stenhouse offered invaluable comments that have been essential in shaping the volume.

We would also like to express our gratitude to the two anonymous readers for precious suggestions and remarks. We owe special thanks to Arjan van Dijk and Ivo Romein from Brill for their incredible support and work for this book.

List of Illustrations

Introduction

Bianca de Divitiis, Fulvio Lenzo, Lorenzo Miletti

In 1514, in the midst of the war of the League of Cambrai (1508–1516) and just a few months after the great fire which destroyed the commercial neighbourhood of Rialto, an elegant folio volume entitled *De Nola* was published in Venice. Divided into three books and enriched by four engravings, *De Nola* is an extraordinary historical, chorographical and topographical treatise celebrating the city of Nola in the Kingdom of Naples. Its author was the Nolan physician and humanist Ambrogio Leone, who dedicated the work to Enrico Orsini count of Nola, while its publisher was Joannes Rubeus, or Giovanni Rosso from Vercelli. The volume would mark an important advance in European humanistic and antiquarian debates. Leone's description of a seemingly minor urban centre in the peninsula is an innovative and ground-breaking work of Renaissance scholarship. Several decades after Biondo Flavio's studies of Rome and Italy had appeared in print, *De Nola* marked a shift in antiquarian publications, and opened the way to a new approach to the description of cities. Its fame was enduring, reaching beyond the borders of the Italian peninsula to the main centres of Renaissance European culture.

Despite its profound originality and its early and widespread circulation, *De Nola* has remained at the margins of Renaissance studies. The essays here collected, therefore, represent the first systematic and comprehensive study of Ambrogio Leone's *De Nola*. Adopting an interdisciplinary perspective, seven contributions by scholars in different fields look at the bibliographical features of Leone's book, his textual and visual sources, both ancient and modern, the artistic and architectural culture it emerged from, as well as the form and function of the engravings. The historical, social and anthropological aspects of the work are also analysed. By considering *De Nola* from all these different points of view and revealing the many elements which contributed to the pioneering nature of Leone's work, this collection of studies tries to restore an important but neglected scholarly achievement of the Renaissance to its rightful place in European cultural, intellectual and artistic history.

1 The Author: Ambrogio Leone

Ambrogio Leone was born around 1458 in Nola, a city of about 4,000 inhabitants, which flourished under the feudal rule of a branch of the Orsini family and enjoyed a strategic commercial and military position within the Kingdom of Naples.[1] Leone spent the first part of his life in his native city, where his family was part of the local mercantile and intellectual elite (his father was a merchant and his uncle was responsible for supervising the major building projects commissioned by the Orsini); he then graduated in medicine probably at the university of Naples.[2] It is known that Leone frequented the Neapolitan humanistic milieu, and became friend with some of the members of Pontano's Academy, including Giovanni Pontano himself and Jacopo Sannazaro. He probably practised medicine in his native city while maintaining close connections with Naples, as well as with other cities outside the Kingdom. An idea of the extended cultural networks that Leone already enjoyed throughout the Italian peninsula at the time comes from the collection of poems that he commissioned from several poets in 1496, to celebrate the sculptural portrait executed by Tommaso Malvito of the young noblewoman from Nola Beatrice de' Notariis.[3] At the beginning of the sixteenth century, Leone moved to Venice. It is not clear exactly why he abandoned his beloved birthplace, but it is not improbable that the enduring state of war and political uncertainty in the Kingdom of Naples played a part in his decision. In Venice he soon became highly valued as a doctor and humanist: he entered into close contact with Aldo Manuzio and the 'Filelleni' circle and studied Greek with Marco Musuro, who became one of his closest friends, together with Aldo himself, Giovan Battista Egnazio and Girolamo Aleandro. In Venice he also befriended Erasmus of Rotterdam, who would long remain his correspondent, and included references to Leone in his works.

In 1508 Leone attended Luca Pacioli's lesson on Euclid held in the church of San Bartolomeo in Rialto. It is probable that he knew some of the other scholars who attended the same lecture, such as the architects Fra Giocondo da Verona, Pietro Lombardo and Giorgio Spavento, and the painter and mapmaker Francesco Rosselli.[4] As we shall see, his participation in such a dynamic milieu encouraged Leone to write and publish several works with Venetian printers. It is noteworthy that his death and funeral held in the same church

---

1   Remondini 1747–1757; Vincenti 1897; Muscarello (ed. Chiarini) 1972; Toscano T. 1996; Avella 1996–1998; Buonaguro 1997; Caianiello 2003; Clarke 2003; Di Cerbo 2013.

2   Croce (1936) 1953; Ammirati 1983; Sica 1983; Defilippis 1991; Spruit 2005; Vecce 2007.

3   Cicognara 1823–1824, IV, 409–412; Percopo 1893; Montera 1934; Castoldi 1989; Castoldi 1992; Bolzoni 2008, 51–53, 157–167; Bolzoni 2010, 166–170.

4   Euclid (ed. Pacioli) 1509; Benzoni 2014.

of San Bartolomeo in 1525 were recorded by Marin Sanudo, who dedicated an encomiastic epitaph to Leone in his *Diaries*.[5]

De Nola, which came out in 1514, was the first of Leone's books to be published. In 1517 he published the *Castigationes adversus Averroem*, a monumental work in thirty books dedicated to Pope Leo X which aimed at refuting Averroist doctrines; a second part came out in 1524 with a dedication to the Doge Andrea Gritti.[6] Just these two dedications suffice to give an idea of the importance of Ambrogio Leone in the international humanistic milieu of his time. In 1519, Leone published a translation of the Greek medical treatise *De Urinis* by Joannes Actuarius, and in 1523 he completed the *Novum opus quaestionum*, an elegant folio volume consisting of 404 questions of an encyclopaedic nature, one of the most important and interesting examples of this genre published in the sixteenth century.[7] Two of his works were printed together just after his death, thanks to his son Camillo: a dialogue on the concept of nobility entitled *De nobilitate rerum* and a Latin translation of the pseudo-Aristotelian treatise *De virtutibus*.[8] Several years after his death, in 1549, his notes on Alexander of Aphrodisias' treatise *De diffinitione* were printed, and again in 1553.[9] All these works reflect Leone's medical training and expertise and his interest in Aristotelian philosophy. *De Nola* was an exception, both as a new genre of antiquarian treatise and in the context of Leone's publishing activity.

2   The Book: *De Nola*

De Nola was printed by the Venetian firm of Giovanni Rosso from Vercelli as an elegant folio volume, enhanced by four engravings produced with the help of Girolamo Mocetto, a Venetian painter belonging to the school of Andrea Mantegna.[10] The title-page has no architectural frame or ornamental borders; instead, the title is followed by a series of adjectives which sound very much like Leone's own description of his work: *Opusculum / distinctum / plenum / clarum / doctum / pulcrum* [sic] / *verum / grave / varium et / utile*. After the

---

5   Sanudo (ed. Berchet et alii) 1879–1902, XXXVIII, 54, 57.
6   Leone 1517; Leone 1524.
7   Leone 1519; Leone 1523.
8   Leone 1525.
9   Leone 1549.
10   Croce (1936) 1953; Spampanato 1921, I, 8–18; Weiss 1958, 178–179; Maiuri 1959; Manzi 1973; Ammirati 1983; Sica 1983; Defilippis 1991; Toscano T. 1996. On Mocetto see Romano 1996; Tagliaferro 2011.

preface, in which Leone dedicates the work to the last count of Nola, Enrico Orsini, and explains the main aims of the book, the contents of *De Nola* are divided into three books, each of which comprises approximately fifteen chapters.[11]

The work begins with a description of the *Ager Nolanus*, the Nolan territory and its borders, where, following Biondo's example, Leone tries to match the ancient sites with present-day ones and, combining material and literary evidence, attempts to describe and reconstruct the city as it was in antiquity. The second book provides a description of Nola in Leone's time, while the third book is a detailed account of the city's families, and of local customs and traditions. The four engravings are closely related to the contents, with each illustration explicating arguments in the text. The first engraving, the only one signed by Mocetto, represents the *Ager Nolanus*, and shows Nola's relation to the surrounding territory and, above all, to Naples, the capital of the Kingdom. The second is a plan of the *Nola Vetus*, based on both material evidence and literary testimonies, with three-dimensional reconstructions of the ancient monuments and an indication of the perimeter of the new city. The third engraving illustrates the geometrical and mathematical procedures by means of which Leone was able to determine the dimensions and perimeters of the ancient city on the basis of the modern one. Finally, the fourth plate is an outstanding image of the Nola of his time, featuring the city's main roads and buildings.

In order to describe his native city, Leone combined different literary genres: historical, antiquarian, chorographic and laudatory. These elements were merged in an unprecedented synthesis, an original and even eccentric combination, which is one of the reasons why *De Nola* has been excluded from specialized studies. The way in which Leone employs different types of sources, both literary and material, ancient and modern, was also new. Although his use of inscriptions and ancient remains finds a parallel in earlier and contemporary antiquarian works, *De Nola* is exceptional for Leone's idea of describing Nola not only in words, but also by giving an essential role to images—a complete innovation in terms of the genre of publication. The volume is, moreover, an instance of the early use of engravings, as opposed to woodcuts. The four elegant copperplates were conceived as works of art, made according to the most advanced technique, introduced by Andrea Mantegna. The richness of the text is in great part due to Leone's wide range of skills: a doctor, imbued with Aristotelian culture, knowledgeable in both Latin and Greek, as well as geometry, natural philosophy and music, an expert in art and architecture, and a patron of the arts, he had both the ability to explain scientific natural

---

11   See Appendix 1 and 2, 163–166.

phenomena, such as the frequent floods that afflicted Nola or the eruptions of Vesuvius, and to describe ancient ruins and modern buildings, such as the city's amphitheatre and the cathedral.

The innovative character of *De Nola* was immediately recognized by Leone's contemporaries. In 1518, Erasmus, having been informed by Leone that his work was being printed, praised his friend's patriotic initiative.[12] Six years later, the Neapolitan humanist Pietro Summonte, in his famous 1524 letter on the state of art in the Kingdom, advised his friend Marcantonio Michiel to read *De Nola*, in which he would find detailed information on the palace in the city where the emperor Augustus had died.[13] In the middle of the sixteenth century, Leandro Alberti invoked Leone's authority in topographical studies;[14] in the last years of the same century the *opusculum* by Leone was indicated as a model by Matteo di Capua, Prince of Conca, to the historian Angelo Tonsi of Fano who wanted to write a history of the city of Caiazzo.[15] The work by Leone continued to be read over the centuries, also thanks to new editions, sometimes with slightly different titles, printed in Italy as well in other European countries.[16] The eighteenth-century abbot Gianstefano Remondini, though very critical toward Leone, whom he blamed for not giving sufficient prominence to Nolan saints and accused of many historical and topographical errors, wrote his *Nolana ecclesiastica historia*, mainly relying, for the earlier period, exactly on what he could read in *De Nola*.[17]

3   *De Nola* in the European Humanistic Debate

While Ambrogio Leone and his connections with other humanists have been the object of some scholarly attention in the last sixty years, *De Nola* has remained largely the preserve of local and folklore studies, when it has not been mined merely to extract particular pieces of information. Despite the brief but significant observations by Roberto Weiss, no attempt has been made to get to

---

12   Erasmus (ed. Allen) 1913, 352–355.
13   Nicolini 1925, 174, 177, 281–282; Pane 1975–1977, I, 63–71.
14   Alberti 1550, 124 *verso*.
15   Pedio 1973, 257–258.
16   Leone 1600; Leone 1614; Leone 1723; Leone 1725; Leone 1729; Leone 1735. Modern Italian translations: Leone (ed. Barbati 1934); Leone (ed. Ruggiero) 1997.
17   Remondini 1747–1757.

grips with the complexity and originality of *De Nola*, and there is still no comprehensive study of Leone's work.[18]

In order to fill this void, our volume aims to shed new light on the varied contents of *De Nola* and to restore a sense of its importance as a substantial contribution to the culture of Renaissance Europe. The essays were originally delivered at two seminars on *De Nola*, held in Nola (April 2012) and in Venice (May 2012), and subsequently developed as part of a research project carried out by the members of the ERC-HistAntArtSI group. In view of the complexity of the subject matter and of Leone's treatise itself, three other scholars were invited to contribute to the volume. Taken as a whole, the seven essays and the bibliographical note look at Leone's work from different points of view with the aim of highlighting the manifold skills of its author and the variety of tasks which were involved in its composition.

De Nola was the first systematic and illustrated antiquarian description of a town outside Rome. Therefore, having no direct models to follow, Leone relied, at a first level, on a massive use of written sources relating to a wide range of disciplines. He appears as a reader (and certainly also an owner) of manuscripts and above all of printed editions, like those published by his friend Aldo and composed in Greek characters with the support of Griffo's refined Greek typeset; but also as a tireless 'devourer' of mathematical and geometrical treatises, like Luca Pacioli's edition of Euclid.[19] He certainly consulted (or even possibly possessed) illustrated books like the famous *Hypnerotomachia Poliphili*, or the 1511 edition of Ptolemy's geographical work (edited by another *émigré* from the Kingdom of Naples, namely Bernardo Silvano of Eboli), or architectural treatises like Fra Giocondo's edition of Vitruvius, all printing products which displayed an impressive iconic apparatus; but he also studied a deliberately aniconic book as Leon Battista Alberti's *De re aedificatoria*, a real bestseller at the time, where mental images were created by the power of words.[20] Ambrogio Leone considered his book not only a medium for the publication of his text, but also a valuable art object, in which refined illustrations play a relevant role. This was just the first of the many features which made *De Nola* markedly different from previous antiquarian publications which followed the model of Biondo Flavio.

---

18   Weiss 1958, 178–179; Weiss 1969, 127–129.

19   Euclid (ed. Pacioli) 1509; Benzoni 2014.

20   Colonna 1499; Ptolemy (ed. Silvano) 1511; Vitruvius (ed. Giocondo) 1511. For Alberti's 'iconophobia', see Carpo 1998; Carpo 2003.

Leone's multiple interests correspond to multiple competences: as his later *Opus quaestionum* also shows,[21] Leone was not only an erudite physiologist, but he had a well-grounded knowledge of mathematics, physics, architecture, engineering, mechanics, geology, botany, and so on; all competences that he allows to emerge in writing his *De Nola*, in which his birthplace is described as a complex object, the clear understanding of which is possible only through a cross-disciplinary method. In this sense, the authors of this volume have 'disassembled' what is, in Leone, a unique and all-comprehensive encyclopaedic approach, the product of a typical humanistic approach which does not hold back from encompassing the whole of human knowledge. In his contribution, Lorenzo Miletti studies the various genres represented in *De Nola* and Leone's debt to several antiquarian and chorographical works of the Renaissance, as well as to the tradition of *laudationes urbium*. He considers Leone's humanistic methods and his use of Greek and Latin sources, together with an examination of the printed editions which he could have consulted while composing his work. Bianca de Divitiis and Fulvio Lenzo investigate the antiquarian method adopted by Leone in describing ancient Nola. Leone mostly relies on literary sources, but he also adds a lot of fresh information derived from the study of epigraphic evidence and of the archaeological remains partly visible and partly buried by the rise in the soilsurface. The essay written by Fulvio Lenzo discusses the four engravings, which help to make Leone's book an early sixteenth century masterpiece of humanist literature. His analysis of the engravings clarifies the important role played by Leone himself in planning and drawing the illustrations, and reduces the role of the Venetian painter and engraver Girolamo Mocetto to little more than an executor of Leone's ideas. Bianca de Divitiis surveys the profound architectural knowledge which is apparent throughout *De Nola*, both in the reconstruction of ancient monuments and in the description of modern buildings. In this context, the chapter on the *domus* is an exceptional account of an abstract model of a Nolan house, based on information found in the treatises of Vitruvius and Alberti. Fernando Loffredo's essay, on the basis of an anecdote recounted by Leone himself in the third book of *De Nola* concerning his commissioning of a marble bust of his beloved Beatrice De Notaris from the Lombard sculptor Tommaso Malvito, looks at Leone's relation to the artistic production of his time and his role as a patron who was aware both of more recent artistic trends and of the dialogue between art and poetry. Giuliana Vitale explains Leone's general purpose in the work as being to preserve knowledge of the nobility of both ancient and contemporary Nola for posterity. By focusing on the third book, which is

---

21   Leone 1523.

devoted to the leading Nolan families, she sets out Leone's idea of nobility and of the social and economic life of the city, as well as its role in the context of the Italian peninsula as a whole. Eugenio Imbriani deals with the last part of the third book of *De Nola*, in which Leone describes the costumes of the inhabit-ants and the rituals, ceremonies and festivities that took place in the city. By adopting an anthropological approach, important rituals, such as the one held at the time of the grape harvest, are analysed in the social context of southern Italy and as a survival of ancient Nolan customs. Finally, the distinctive place of *De Nola* in the context of contemporary printing and publishing in Venice and elsewhere at the beginning of the sixteenth century is examined by Stephen Parkin, drawing on techniques of bibliographical analysis and description. The passages from the *De Nola* which are more extensively discussed in the essays feature as appendixes at the end of the volume.

To sum up, the contributors' effort to analyse different aspects of *De Nola*, considered together, gives a complete picture of the context, aims and out-come of Leone's book. Taking into account the complexity and originality of the work, they explore important aspects which have never been considered before. The fact that a book written by a Nolan humanist and devoted to a town of southern Italy was published in Venice in itself brings new light on how the cultural connections between the Kingdom of Naples and other states of the peninsula were deeper than is commonly recognized. In this perspec-tive, the case of Leone is particularly relevant to understanding the circulation of humanists and ideas within the Italian peninsula and across different coun-tries of Europe in the early sixteenth century. Compared to Biondo and his early followers in antiquarian studies, Leone belonged to a new generation of scholars who, driven by political and religious conflict or impelled by scholarly research and curiosity about natural phenomena, travelled extensively across Italy and Europe, with their paths continually crossing, and who belonged to interconnected circles, like those of Giovanni Pontano in Naples and of Aldo Manuzio in Venice.

The geographical crossing between southern and northern Italy, underlying a book printed in Venice but dealing with a city which belonged to the Kingdom of Naples, was due not only in general to Leone's travels and scholarly con-nections, but also to a quite specific circumstance: at the turn of the fifteenth century, Nola was famous in Venice as the principal fiefdom of a national hero, Niccolò Orsini, commander in chief of the Venetian army from 1496 and a key figure in the War of the League of Cambrai, who was celebrated after his death in 1510 with a monumental tomb in the church of Santi Giovanni e Paolo.[22]

---

22   See Miletti and Loffredo in this volume, respectively 12–14 and 115–120.

In those years Venice would therefore have been receptive to Leone's wish to celebrate Nola and to preserve for posterity the memory of the nobility of his native city. A work on the southern capital of the Orsini family's possessions would have aroused interest not only among the ruling elite but very likely also among a wider audience. Furthermore, that this antiquarian description of Nola was printed as an expensive and finely illustrated printing product, testifies how the whole Venetian milieu was interested in southern Italian matters, a fact which was clearly confirmed, a few years later, by the monumental Aldine edition of Giovanni Pontano's *Opera* published between 1518 and 1519.[23]

By analysing Leone's literary masterpiece from different points of view, a single element emerges as characteristic, at a deeper level, of the whole work, independently from its multifaceted essence, namely Leone's sense of the past. His perspective is no doubt an antiquarian one, but the pride of place he clearly gives (not unlike every humanist of his time) to classical antiquity does not imply that later periods can be disregarded nor that present times should be condemned. Antiquity itself is not seen as a monolithic object, rather as a stratified one, which, as in the case of Nola, it is possible to divide into various phases—Italic, Greek, Samnite, and Roman—all of them worthy of our consideration, in so far as all of them contributed to the definition of the city's identity. Such an approach, which is indebted to earlier antiquarian works like Biondo's *Roma instaurata* or Fra Giocondo's prefatory epistle to his collection of inscriptions addressed to Lorenzo the Magnificent, had a meaningful parallel in about the same years in Raphael's epistle to Leo x.[24] Rather than stressing a historical continuum, however, Leone aims to match past with present by conceiving a thematic development for his work, as is clear from the accurate index he inserts at the beginning of the book.[25] He devotes no chapter specifically to a 'history of Nola'. A systematic overview of the city's history is simply missing. Throughout the whole work, he reports and celebrates several deeds dating back to the Roman period, and discusses, by mentioning St. Felix and St. Paulinus, selected fourth-century CE episodes. He next refers to episodes dating to the late Anjou and Aragonese periods. Thus, the long period running from late antiquity to the early Anjou period is almost entirely omitted. It is probable that such an omission reflects, more than a purely ideological motivation, a comprehensive lack of relevant sources for this period: apparently, Leone could not entirely follow his model Biondo, who gave great emphasis to

---

23   Pontano 1518–1519.
24   See Frommel, Ray, Tafuri (1984) 2002, pp. 396–399, 437–450; Rowland 1994; Koortbojian 2002; Biondo (ed. Raffarin-Dupuis) 2005–2012.
25   See Appendix 1, 163–164.

the Middle Ages in writing his *Decades*, nor his beloved Giovanni Pontano, who displayed a certain interest in the Norman period of the Kingdom of Naples.[26] Be that as it may, it is a fact that Leone's exploration of written and material sources in search of the city's past does not coincide with (nor appears to aim for) a reconstruction of local history pursued through a diachronic approach. If not purely historical, Leone's antiquarianism has above all a moral purpose: his main desire is to demonstrate that the Nolan citizens of his times are animated by the same virtuous spirit which inspired their behaviour, for instance, during the second Punic War, when they defeated Hannibal's troops. Thus, the reconstruction of Nola's past is functional to understanding Nola's present. The virtues of modern Nolan citizens, as well as the architectural beauties of modern Nola find their homologues and their harbingers in the spotless *mores* and in the ancient monuments of the Roman period. Where it was impossible to construct the historical continuum, Leone built a 'moral' and 'metahistorical' bridge from antiquity to the present.

Taken as a whole, the selected aspects of Leone's complex method which we have tried to untangle in this volume encourage us to see *De Nola* within the broad picture of European humanism, recognizing the central role of Ambrogio Leone's book in the intellectual debate of his time, and also to rethink those deeply rooted hierarchies, which still relegate southern Italy at the margins of Renaissance studies. If a new edition of *De Nola*, with a philological revision of the original Latin text and a translation into English is desirable, we hope that in the meantime readers of our book will be encouraged to seek out Leone's oeuvre, and possibly even take a copy of this elegant fifteenth-century volume in their own hands.

---

26   On Pontano as interpreter of the medieval history of southern Italy in his *De bello Neapolitano* see Miletti 2018; de Divitiis forthcoming.

CHAPTER 1

Ambrogio Leone's *De Nola* as a Renaissance Work: Purposes, Structure, Genre, and Sources

Lorenzo Miletti

Printed in Venice in 1514 but composed for the most part in 1512,[1] Ambrogio Leone's *De Nola* constitutes a pioneering and experimental work, a complex textual object which crosses different genres and encompasses a remarkable variety of methods, arguments, and sources.

The aim of this chapter is to illustrate the characteristics of the *De Nola* as a literary work, focusing in particular on its purpose, structure, genre, and sources. These topics are discussed separately, in four different sections: in the first, the purposes of the *De Nola* are discussed by analysing Leone's preface to the work; in section two, a full description of structure and contents is provided; section three deals with the problem of this work's genre; in the last section, Leone's use of Greek and Latin sources is examined.

1　The Title and the *Praefatio*: History and Rhetoric

The title *De Nola*, by which the work is generally known, is the shorter form of the longer title printed on the title-page:

> DE NOLA | opusculum | distinctum | plenum | clarum | doctum | pulcrum [*sic*] | verum | grave | varium et | utile

This 'little work' (*opusculum*) is defined with no fewer than nine attributes inviting the reader to concentrate on its literary and moral qualities. The adoption of a *de* + ablative clause reveals the author's intention that his work should stand among scientific and descriptive treatises, as he explicitly claims in the preface.

The work's contents are distributed into a preface (*Praefatio*) and three books. The *Praefatio* (whose complete text is transcribed below, in

---

1　See the incipit of book 2, chapter 1: "Quonam autem modo nunc se habeat, hoc est anno MDXII post Iesum Dominum, consequens est enarremus" (Leone 1514, f. xxii *recto*).

Appendix 2) opens with a dedication which contains for the first time in the volume the author's name, absent from the title page, and the name of the dedicatee, the Count of Nola Enrico Orsini:

AMBROSII LEONIS IN LIBELLOS DE NOLA | PATRIA AD ENRICVM
VR|SINVM PRINCIPEM IV|STISSIMVM

The phrase which refers to the book is slightly different from that printed on the title page: here we find *libellos de Nola patria*, instead of *de Nola opusculum*. Both *libellus* and *opusculum* reflect a wish to understate, which is clearly in contrast with the book's folio format and its not inconsiderable length.[2] As for the presence of the word *patria*, Ambrogio also refers to his work as *De Nola urbe patria nostra* in an epistle to his friend Erasmus of Rotterdam: such an expression clearly implies that the book has an encomiastic and patriotic purpose.[3]

The work's dedicatee, Enrico (or Arrigo) Orsini, was the last count of Nola, who died in 1533 after having taken part in a revolt against the emperor Charles V and, as a consequence, falling into disgrace.[4] After Enrico's death, Nola was 'freed' from its condition as a feudal city and was included in the *regio demanio*, *i.e.* it came under the direct rule of the King. When the *De Nola* was published, however, Enrico's position was still favourable towards the emperor and the Neapolitan court. Enrico was the grandson of Niccolò Orsini, Count of Nola and Pitigliano and commander in chief of the Venetian army from 1495 to his death in 1510 [fig. 62].[5] Thus, at the beginning of the 16th century, Nola had become much better known in Venice thanks to the positive role played by Niccolò in the War of the League of Cambrai, despite his (debatable) responsibility in the disaster which befell the Venetian army at Agnadello against the French in 1509. Although, after this battle, the Venetian Republic, which had lost almost all its mainland territories, faced a crisis, Niccolò was successful in turning the situation around by using a stratagem to retake Padua and putting

---

2 For a bibliographical description of the book see Parkin in this volume.

3 Erasmus (ed. Allen) 1913, 355 epistle 854.76–85: "Opusculum enim in tris distinctum libros de Nola urbe patria nostra composuimus, quod impressum est". For an analysis of the correspondence between the two humanists see Vecce 2000; Miletti 2016a.

4 A biographical profile of Enrico Orsini can be found in Vincenti 1897, 63–72.

5 Oddly, the *Dizionario Biografico degli Italiani* includes no entry for this important *condottiero*; for a biography see Bruscalupi 1872; Vincenti 1897, 41–64; Norwich 1983, 399–406. A relevant source for his life is still the speech written on the occasion of his funeral by the humanist Giovan Battista Egnazio (Egnazio 1510).

up a strong resistance to a fresh siege mounted by the League's army. After his death, he was celebrated as a national hero, and a monumental sepulchre was built in his honour in Venice, in the church of Santi Giovanni e Paolo [fig. 64]. In the light of Niccolò's role in Venice, Leone's dedication to Enrico can probably be understood also as a homage to the most direct contact between Nola and Venice in that period, namely the Orsini family.[6]

The text of the *Praefatio* is a useful tool with which to understand the author's purposes. It begins with a clearly intelligible allusion to the *incipit* of Sallust's *The Conspiracy of Catiline*, via the rhetorical *topos* according to which an individual may benefit his homeland not only with deeds, but also with the fruits of his *ingenium*, i.e. his intellectual activity:

> Every man, Prince Enrico, keenly desires with extraordinary striving to benefit his homeland with his utmost strength and to honour her with every service, veneration and piety.[7]

By evoking Sallust, Leone aims to bestow on his work the moral purposes which were characteristic of classical historiography. The '*bene facere rei publicae* theme' (Sallust *Cat.* 3) dominates the first section of the preface: a few lines later, Ambrogio mentions two men who through their deeds enhanced the greatness of their homelands. The first example he gives, the Roman *pater patriae* Furius Camillus, is drawn from Antiquity, the second, Alfonso II of Aragon, king of Naples (1494–1495), from the very recent past:

> For this reason, with ardent spirit, with commitment and care, Furius Camillus took revenge for the devastation of Rome, which had been burnt down and destroyed by the Gauls, and restored and enlarged it. Similarly, Alfonsus II of Aragon, the brother of your grandfather, enlarged his hometown Parthenope, which previously was almost left open [to attack] and was contracted [in size], and then enclosed it with extremely strong walls.[8]

---

6 For a discussion of the Venetian milieu which included both Niccolò Orsini and Ambrogio Leone see also Loffredo in this volume, 115–121.

7 Leone 1514, f. ii *recto* (Appndix 1, §1): "Omnes homines Enrice Princeps vehementer optant mirificeque nituntur ut patriam tam ope summa iuvent quam omni officio, veneratione ac pietate prosequantur". Cf. Sallust, *De Catilinae coniuratione* 1: *Omnis homines qui sese student praestare ceteris animalibus summa ope niti decet*, etc.

8 Leone 1514, f. ii *recto* (see below, Appendix 2, § 3): "Eamobrem combustam dirutamque a Gallis Romam Furius Camillus ardenti animo, cura, sollicitudine ab interitu vendicavit,

Camillus, whose reputation in the Renaissance stemmed from knowledge of Plutarch's life of him, was a recurrent example of *defensor patriae*.[9] As a hero of the Roman republic, he was also well suited to the 'republican' tastes of Leone's Venetian readers.[10] But it is possible to infer something more from Leone's allusion to Camillus: the solemn sentence quoted by Leone, which literally refers to the Roman general, also seems to evoke, covertly, the personality of Niccolò Orsini mentioned above. Niccolò's desperate (and successful) re-conquest and defence of Padua and of the mainland Venetian territories were comparable to Furius Camillus' desperate recapture of Rome against the Gauls.[11] Furthermore, the hostile mention of the *Galli* in Leone's words ("combustam dirutamque a Gallis") appears to be an allusion to the army which defeated Venice at Agnadello, as seen above.[12]

As for Alfonso II of Aragon, his presence alongside such a celebrated hero of Antiquity as Camillus might at first sight seem surprising. An initial reason for his inclusion in the *Praefatio* is the fact that Alfonso's brother, Enrico of Aragon (c. 1450–1478), was the father of Caterina, wife of Gentile Orsini and

---

instauravit, adauxit. Alfonsus item rex Aragoneus secundus, Enrici avi tui frater, Parthenopen patriam, quae iampridem aperta prope atque angusta fuerat, tum plurimum amplificavit tum moenibus quam praestantissimis circundedit".

9   On the reception of Plutarch's *Vitae* in the Renaissance see Pade 2007 (Camillus: 47–56). As for references to this episode in Renaissance art, see, for instance, Francesco Salviati's frescoes in Florence (1543–1545), in the Sala delle Udienze di Palazzo Vecchio, which are entirely dedicated to an account of the actions of Camillus. On this subject see Caciorgna, Filippini, Guerrini 2001.

10   A certain preference for the republican history of Rome is detectable in the whole work. In order to celebrate Nola's past, for instance, Leone takes pains to stress more than once the bravery of the Nolans against Hannibal during the Second Punic War, while he does not dwell too much on the fact that the city was the place where Augustus died; on these topics see Miletti, 2016b.

11   See also the words of one of Ambrogio Leone's closest friends, Battista Egnazio, who, in his funeral speech in honour of Niccolò Orsini, praised him by comparing—by means of the rhetorical device of *praeteritio*—his deeds to those of several famous Roman republican heroes, including Furius Camillus: "... non video quem veterum ducum illi [*scil.* Nicolao] anteferre, nostrorum conferre valeas, aut cuius res gestae tantae talesque esse possint, quae cum huius gloria et amplitudine committi compararique debeant. Atque ut illa vetera Reipublicae Romanae lumina Camillos, Fabios, Scipiones, Caesares omiserim ...". (Egnazio 1510, f. [b vi] *recto*).

12   A similar perspective is adopted in the whole of Egnazio's funeral speech in honour of Niccolò (see the previous note), in which we find several references to the battles waged by Niccolò against the French (*Galli*), beginning with Charles VIII's campaign of Italy (1495).

mother of the dedicatee of the *De Nola* Enrico, as Leone makes clear through his use of the expression "your grandfather's brother" ("avi tui frater").[13] It is also possible that Leone, who had lived in the Kingdom of Naples under Alfonso II, was deeply attached to the memory of his former sovereign. Yet the reference to Alfonso should not be interpreted as a merely courtly homage, nor as a symptom of the author's homesickness: in Leone's opinion, Alfonso constitutes the outstanding example of a sovereign who was able to accomplish important civic deeds. His presence in the passage therefore balances the reference to the solely military deeds accomplished by Furius Camillus. Alfonso intervened on several occasions in the urban planning of Naples, and provided the city with new walls.[14] Alfonso's restoration of the city walls in Naples attracted attention between the end of the 15th and the beginning of the 16th century: the king's interventions were described and praised by Giovanni Pontano, by Bernardo Rucellai, and also by Marin Sanudo of Venice, who mentions them in his work on the Italian campaign of Charles VIII of France.[15] Furthermore, as we shall see, Leone was extremely sensitive to urban and architectural issues, and this sensibility may have played a role in his wish to stress Alfonso's achievements.[16]

In the immediately following sentence, Leone cites two ancient Greek writers, Isocrates (5th–4th century BCE) and Aelius Aristides (2nd century CE), who celebrated and glorified Athens by composing Panathenaic orations.[17] The joint reference to these two authors is probably due to the fact that Aldus Manutius had recently published, in 1513, an edition of both Isocrates' and Aristides' *Panathenaici* in a volume devoted to the *Oratores Graeci*, the prefatory epistle

---

13   Caterina was the daughter of Enrico of Aragon, Alfonso II's half-brother, since he was the son of Ferrante I king of Naples and Giovanna Caracciolo. Enrico was baron of Gerace, a town in southern Calabria, and died in 1478 from mushroom poisoning, when he was only twenty-seven years old. His tragic death was mourned in a vernacular poem, written by Joanne Maurello and printed in Cosenza in the same year as Enrico's death (Maurello 1478).

14   This restoration is felt by Leone to be particularly important: for the depiction of the restored walls in Naples in the *De Nola*, in Mocetto's engraving entitled *Ager Nolanus*, see Lenzo in this volume, 65. In the index of contents, furthermore, Leone inserts Alfonso's deed among the topics treated in his *Praefatio* (see below, Appendix 2, § 3), while he does not mention the remaining examples.

15   Cf. de Divitiis 2011, 324–330.

16   Cf. de Divitiis, Miletti 2016; see also both Lenzo and de Divitiis in this volume.

17   Isocrates, or. 12; Aelius Aristides, or. 1.

of which was addressed to Leone's friend Giovan Battista Egnazio.[18] The date of this *editio Aldina*, printed while Leone was working on the *De Nola*, and the close friendship between Aldus and Ambrogio, can lead us to suppose that the latter was probably well acquainted with the former's interest in Isocrates and Aristides. The mention of the two Greek orators may also be read as an allusion to some modern encomiastic works which were inspired by these classical models, such as Leonardo Bruni's famous *Laudatio Florentinae urbis*, which was explicitly inspired by Aristides' *Panathenaic*, a text known to Bruni thanks to his Greek teacher Manuel Chrysoloras.[19]

In the rest of the passage Leone states that he aims to present all these grand personalities as 'other than self', adding that he too intends to benefit his *patria*, but his contribution will seem modest when compared with those of the men he has just mentioned:

> I too intend to honour my homeland with my efforts, just as I seem to emulate all these men's desire to do good, so the service I am going to offer is somehow different. While they restored or enlarged their homelands with great works, I have made her visible, with help from the painter Girolamo Mocetto, with an image which replicates her form so exactly that she can very easily be viewed by the eyes of all men in every part of the world. Again, while they praised their homeland and exalted it to the heavens elegantly and exhaustively, I have done this, on the contrary, by embracing the simplicity itself of history, setting aside copious eloquence and omitting the ornaments and flatteries of rhetoric. I have decided to honour my country in this little work with a simple image and a purified history.[20]

---

18   *Oratores Graeci* 1513. The preface to this edition has been published and translated into English in Manutius (ed. Wilson) 2016, 222–225.

19   See below, in the third section of the present contribution.

20   Leone 1514, f. ii *verso* (see below, Appendix 2, § 5): "Ipse vero patriam quoque studio prosecuturus, ut horum omnium bene faciendi cupiditatem videor imitari, ita modo quodam sum dispar munere ipso reddendo. Illi enim patriam opere magno redintegravere vel effecere maiorem, ipse imagine atque aspectus proprii similitudine tanta produxi in medium, adiutus opera Hieronymi Moceti pictoris, ut oculis omnium atque ubique terrarum perquam facile possit esse conspicua. Quinetiam illi patriam concinne affatimque laudaverunt ac evexerunt ad astra, ipse contra copia dicendi seposita relictisque floribus et orationis lenociniis sola sum historiae simplicitate complexus. Imagine namque pura atque historia defaecata in hoc opusculo decrevi patriam venerari".

If those personalities made their country greater thanks to their great works (*opere magno*), he, Leone, will make Nola known by making it visible to "the eyes of everyone" ("oculis omnium ... possit esse conspicua"), thanks to the help of the Venetian painter Girolamo Mocetto, whose engravings are included in the volume [fig. 1–4]. While Isocrates and Aristides praised Athens with the power of their eloquence, he will renounce all rhetorical 'flowers' and embrace the sole *simplicitas historiae, i.e.* in straightforwardly descriptive prose focusing on real information only.

It is of great interest to see how Leone stresses Mocetto's contribution, and how he shows himself aware of the pioneering role played by the four engravings.[21]

Interesting too is the polemic against rhetoric, conducted—paradoxically—with very 'rhetorical' arguments, namely by opposing the presumed simple and 'objective' character of historiography to the excesses and flatteries of rhetoric. Nevertheless, this polemic allows us to understand that Leone is eager to classify his work (more or less correctly) as historiography, and is concerned about the possibility that his readers may interpret the whole book as a mere *laudatio*. We will touch again on this issue below, in analysing the problem of *De Nola*'s genre.

The engravings—Leone continues after the passage quoted above—together with the book's historiographical approach make the *De Nola* useful both for its present day readers and for posterity: the iconic power of the images can represent absent objects as though they were present, while historiography can 'make deeds brighter', and in itself brings a "great promise of eternity" ("promissio magna quaedam aeternitatis", see Appendix 2).

In the last part of the preface, Leone explains the reasons why his work could be useful for both Nolans and those who are not natives of the city, since the latter do not know the place at all or only have a vague notion of what it is like, while the former are usually ignorant of their own history and unaware of their own city's merits.[22]

To summarise the topics touched on in the *Praefatio*, this part of the book—which was presumably written after the main body of the work had been finished—reveals the author's main purposes and concerns. The long introductory discussion on patriotic benefactors aims to provide the work with a moral intent. The choice of examples and the arguments adopted are intended to prevent readers from perceiving the book as a merely rhetorical production: while Leone cannot deny that the purpose of his work is also to glorify Nola

---

21   Cf. Lenzo in this volume.
22   Leone 1514, f. ii *verso*.

and is, of course, aware that several chapters of the book have a purely rhetorical shape and structure (as we shall see), he nevertheless aims to persuade his
readers that such an encomiastic effect will emerge from his sole description
of the facts.

2   An Outline of Structure and Content

The distribution of the material into three books corresponds to a tripartite
approach to the city: the first book is about the Nolan territory and, above all,
about the ancient form of the city. The second book is devoted to modern Nola,
with a description of the city and its monuments (churches, walls, palaces
etc.); the third book is a description of Nola's social, economic, religious life
and its popular culture. Each book of the *De Nola* is divided into chapters, each
of which has a specific title, as the author himself scrupulously summarizes in
a detailed table of contents placed at the end of the work, not included in the
pagination.[23]

Since the work still lacks a well-known modern edition and English translation, I give here a summary of the contents of the *De Nola*, with a commentary.

2.1   *Liber primus* (ff. iiii *recto* to xxi *verso*)

In the first book Leone describes the morphology of the Nolan region and the
city's geographical position, discusses some episodes of Nolan ancient history, offers a reconstruction of how Nola was like in antiquity, and finally provides an overview of the main virtues shown by Nolan citizens throughout the
centuries.

The book is composed of fifteen chapters. In the first two, the author describes the territory of Nola, placed between Mount Vesuvius and Mount
Avella, its boundaries, the fertility of its soil, and other characteristics, also
making several remarks on plants and questions of cultivation. Connected to
these chapters is Girolamo Mocetto's first engraving, entitled *Ager Nolanus*,
which represents the Campanian area from Naples to Mount Vesuvius, with
Nola at its centre [fig. 1]. In his description, Leone also refers—as is usual in humanistic writers—to ancient Greek and Latin literary sources to explain and
examine natural phenomena which were still visible in his own time. More
generally, a great number of ancient authors are quoted throughout the entire book: chapter 2, for example, which is devoted to the quality of the soil,

---

23   *Index omnium eorum quae passim in hoc opere continentur.* The Latin titles of the chapters
      are listed below, in Appendix 1.

is for the most part occupied by the discussion of two Virgilian passages, one from the *Georgics* and one from the *Aeneid*, which in the ancient commentaries were both considered to be evidence for Virgil's hostility towards Nola.[24] According to these sources, the name of Nola, which never appears in Virgil's *opera*, was cited, in a first draft, in these passages, but subsequently erased by the poet himself, since the Nolans did not allow him to bring some publicly owned water into a farm belonging to him, situated close to the city. Thus, Virgil's failure to mention Nola was due to the poet's wish to 'take revenge' on the city. Leone's position on this question is, as one might expect, apologetic: he dubs these anecdotes maliciously anti-Nolan, and denies that so sensitive and virtuous man as Virgil could ever have been so petty as to erase the name of Nola for such a trivial reason.[25]

Chapters 3 to 6 concern the origins of the city and its more ancient history, when Nola was allied to the Samnites, before being subjected to Roman power. As for the foundation of the city, Leone attempts, following the accounts of Silius Italicus and Pompaeus Trogus, to date it back to a Greek, namely Chalcidian, colonization.[26] As modern scholarship has shown, the Nolan foundation myth 'promoting' a Chalcidian identity can probably be connected with the influence of the Chalcidian city of Neapolis, since the two centres were linked by solid political and economic relationships before the Samnites took control of the whole Campanian area.[27] Be this as it may, Leone uses this version in order to 'ennoble' his native city and its territory by pointing to its (presumed) Greek origins.[28]

---

24 The Virgilian passages are *Georgics* 2.224–225; *Aeneid* 7.740; the anecdotes about the "author variant" are mentioned by Aulus Gellius, *Noctes Atticae* 6.20, and Servius *Georg.* 2.224 (Servius [ed. Thilo-Hagen] 1878–1887, III.1.240); *Aen.* 7.740 (Servius [ed. Thilo-Hagen] 1878–1887, II.190–191).

25 A full discussion of the anecdotes concerning Virgil and the Nolan water is in Miletti 2016a. See also Lenzo in this volume, 66.

26 Silius Italicus, *Poenica* 12.161: *hinc ad Chalcidicam transfert citus agmina Nolam*; Trogus *apud* Justin 20.1.13: *Iam Falisci, Nolani, Abellani nonne Chalcidensium coloni sunt?* On the Nolan trend, in antiquity, to absorb the Greek identity of Neapolis see Miletti 2014, 36.

27 Cassola 1986, 75.

28 This tendency to 'elaborate' a Greek identity for his native city is something shared by Leone with other humanists of his time who were active in the Kingdom of Naples; see e.g. Antonio Galateo, who, in his *De situ Iapygiae* (for a bibliography see below, in the present chapter), points to the Greek and pre-Greek origins of the Salentine towns, understating the Roman phases of civilization; or Francesco Peto of Fondi, who even attributes Laconian origins to his hometown (see Miletti 2015 and Miletti forthcoming).

Leone discusses quite accurately how and when Nola was allied to the Samnites, also analysing the beginning of the Roman era, *i.e.* the end of the Samnite Wars. At the beginning of chapter 6 Leone uses a vivid expression to describe the pre-Roman period, defining it as the "city's childhood" (*veluti infantiam Nolae urbis*). Despite this awareness of the importance of the Roman phase, a real history of Nola under the Romans is not provided. Great emphasis is laid on single significant episodes connected to the Roman past, but there is no systematic historical narration.

An episode reported by Leone with patriotic enthusiasm in chapter six is the battle of Nola which took place in 215 B.C.E., during the Second Punic War, known above all through Livy's account.[29] After the battle of Cannae, Hannibal occupied almost the whole of Campania, after establishing his headquarters in Capua. He also besieged Nola, defended by the Roman consul Marcellus, who made a sortie with Nolan troops and was successful in causing Hannibal's retreat. The episode was of resounding significance, since the Romans had hitherto been systematically defeated by the Carthaginians.[30] Leone presents it not only as a proof of the Nolans' loyalty to their friends and allies, but also of their military skills. With the same purpose he cites the episode of Erennius Bassus, a Nolan officer loyal to Rome, who refused a proposal of alliance from the Carthaginians (Livy 23.44).

Leone's interest in such episodes is more moral than historical, although his examination of the sources is scrupulous: it is clear from his careful use of Greek and Roman materials that he has a solid awareness of the historical *continuum*; his method, nevertheless, endorses a different approach, closer to an antiquarian and moral/philosophical perception of the past.

Thus, if a systematically 'diachronic' point of view is lacking, Leone's antiquarian sensitivity makes a 'synchronic' approach to ancient Nola possible. Chapters 7 to 8 represent the antiquarian core of the whole book. Here Leone concentrates all his efforts on reconstructing the city as it was in Antiquity. He describes all the remaining vestiges of Antiquity which he could see, and tries to interpret them with a greater or lesser degree of plausibility and convincingness. It is probable that he sometimes makes use of forgery in order to prove or strengthen his point.[31] Following the examples of most Renaissance antiquar-

---

29   Livy 23.16.

30   Cf. Livy's opinion on the significance of this minor skirmish, 23.16.16: *nescio an maxima illo bello gesta sit; non vinci enim ab Hannibale* [*vincentibus*] *difficilius fuit quam postea vincere.*

31   See, for instance, bk. I, ch. 8 (f. xiii *recto*): "In media duorum amphitheatrorum distantia etiam inuenta sunt marmora aliquot quadrata, inter quae unum erat incisum litteris

ian literature, he uses a method which combines autopsy and classical literary sources. The city plan entitled *Nola vetus* [fig. 2] constitutes a 'translation' in graphic terms of all the data reported and analysed in these chapters.

A revealing example of Leone's antiquarian method is his reconstruction of the city walls' perimeter. He infers from Silius 12.162–163 and from the Livian passage quoted above—both concerning the battle won by Marcellus—that the circuit of the city walls was perfectly circular or, as a possible alternative, had the form of a regular polygon composed of six or twelve sides. His hypothesis is based on a rather forced interpretation of Silius's verses, *campo Nola sedet crebris circumdata in orbem | turribus*, where Leone takes *in orbem* to refer to an actual circular shape, whereas in effect it seems to refer generically to the perimeter ("all around"). He finds in Livy 23.16.8, furthermore, the report that Marcellus' troops made their sortie coming out from three city gates.[32] Since, in Leone's opinion, these three gates must have been located in sequence on the same side of the perimeter, he finds it possible to draw the conclusion that there were a maximum of twelve city gates, by calculating three gates in each portion of the perimeter (one quarter of the whole), whether the perimeter forms a circle or a polygon.[33] But Livy's words do not necessarily give rise to such an interpretation, since they do not imply that the gates were exactly on the same side, nor that such a 'side' corresponded to a quarter of a circle, although Leone is right in concluding from the rest of Livy's account (23.16.8–14) that Hannibal's troops were not deployed all around the walls.[34] As one can see, Leone tends to force the interpretation of his sources for the sake of

---

dicentibus TEMPLVM AVGVSTI, caeterae deletae erant, quae certo argumento sunt illic extitisse Augusti templum". Here Leone probably invents the existence of the inscription in order to strengthen his own hypothesis according to which the temple of Augustus was situated in this place. The suspicion that he invented it arises because there is no further reference to this inscription, which was considered to be a forgery by Mommsen (CIL, X, 174*), and also because of its function as a 'providential sign' in Leone's account where it 'confirms' that the temple of Augustus was exactly where Leone wanted it to be. Full discussion of this passage in Miletti 2016b.

32 Livy 23.16.8: *Ad tres portas in hostes versas tripertito exercitum instruxit; impedimenta subsequi iussit, calones lixasque et invalidos milites vallum ferre. Media porta robora legionum et Romanos equites, duabus circa portis novos milites levemque armaturam ac sociorum equites statuit.*

33 Leone 1514, bk. I, ch. 7, f. xi *verso*.

34 On the contrary, in a second battle described by Livy in 23.44, it is clearly stated that the Carthaginians were stationed all around the Nolan walls (23.44.3: *[Hannibal] corona oppidum circumdedit*).

establishing a geometrical symmetry and in order to depict ancient Nola as a kind of ideal city, harmonious in its urban proportions.[35]

As a corollary to his discussion of the ancient urban structure of Nola, Leone goes on to examine the system of canals which had been built to protect the city from the frequent floods, a feat of engineering which no longer existed in Leone's day and the disappearance of which he regrets (ch. 9). In chapters 10 and 11 we find a discussion on the extent of the territory ruled by the Nolans in Antiquity, and on the ancient city of Stabiae, on the coast, which corresponds to present day Castellammare di Stabia but which Leone wrongly identifies with Torre Annunziata.

The last part of book 1 marks a shift to an even more heightened rhetoric. Chapters 12 to 15 are wholly devoted to praising Nola, and, more specifically, the virtues shown by the Nolans from Antiquity until modern times. Leone argues that God has been benevolent towards the city, interpreting as signs of divine favour some characteristics of its location or some episodes of its history. Chapter 13 is an encomium of fidelity and loyalty shown by Nolans towards their rulers throughout the centuries. Examples of such a characteristically Nolan virtue are drawn from antiquity—citing again Erennius Bassus's response to Hannibal quoted above—, and from more recent times: Leone mentions more than one episode in which the citizens showed themselves loyal to the Nolan barons, the Orsini family, even if he does not explain explicitly their political role in the city. Chapter 14 contains a subtle argument beginning with the assertion that in every well-ruled city ("in omni recte constituta et ordine fundata respublica", f. xviii verso) the populace should take care of five aspects of public life, *i.e.* religion, philosophy, military art, agriculture and commerce, and proceeds to demonstrate that Nola did not neglect any of these areas of activity. In this chapter, one of the most patriotic and encomiastic in the entire work, Leone attempts to show the continuity between the ancient pagan *pietas* towards the Olympian gods, and Christian piety, a continuity also testified to by the fact that many ancient pagan temples were transformed into Christian churches, as happened to the temple of Jupiter, on the ruins of which the cathedral and the archbishop's palace were later built. Here Leone also inserts a description of the churches of Cimitile, although he is not aware that it was the place where such Nolan saints as St. Felix and St. Paulinus were buried (as we shall also see below in dealing with book 2, ch. 12). As for philosophy, Leone compensates for the lack of famous Nolan philosophers and writers, whether ancient or modern, by inferring solely on the basis of the fact that the city had its own senate in Antiquity that its citizens were careful to

---

35   See also Lenzo in this volume.

cultivate culture and education, without which no government could survive. More interesting is Leone's extensive use of Roman inscriptions in order to demonstrate the conspicuous presence in Nola of soldiers and officers—and so the city's cultivation of the *ars militaris*—in the Roman period.[36] In chapter 15, the last in book 1, the praise of the city is heightened even further with an argument which aims to stress the Nolans' *elegantia*, a word which would seem intentionally to evoke Lorenzo Valla's work *Elegantiae*, since Valla is cited by Leone as an authoritative source. According to an anecdote reported by Leone, Valla used to say that the Nolans of his times had the best Latin pronunciation of any place in Italy.[37]

From this analysis of the first book two elements clearly emerge, which have not been stressed in the preface and are not explicit in the title of the work: for the sake of simplicity we might label them antiquarianism and natural philosophy. The description and the reconstruction of ancient Nola is the product of the author's profoundly antiquarian studies, while the numerous remarks on the properties of plants, the characteristics of the soil, on watercourses, all spring from Leone's competence in medicine and in natural philosophy, a competence developed thanks to his Aristotelian studies which led him, shortly after the publication of the *De Nola*, to publish a work against Averroism and a collection of *Problemata*.[38] We will find these elements also in the rest of the work.

2.2 Liber secundus (ff. xxii recto to xxxx verso)

The second book is entirely devoted to modern Nola. Not without a certain parallelism to the structure and the arrangement of the arguments found in book 1, Leone describes in a first section the modern *ager Nolanus* (ch. 2 to 6), and in a second, longer, section (ch. 7 to 17) the city itself; his description corresponds to Mocetto's engraving entitled *Nola praesens* [fig. 4], a graphic representation which combines in some sense the features of a plan and an elevation.

In chapter 1 Leone proudly states that modern Nola is located in exactly the same site as the ancient city: the Nolans never moved to more naturally

---

36 In order of appearance: CIL, X, 1273; 1262 and 1258. See the analysis in de Divitiis and Lenzo, in this volume.

37 Leone 1514, bk. I, ch. 15, f. xxi *verso*.

38 Leone 1517 and Leone 1523. It is probable that Leone's interest in Aristotle was further encouraged, in Venice, by the fact that Marco Musuro himself had made a comprehensive study of the Greek philosopher, editing Alexander of Aphrodisias' commentary to Aristotle's *Topica* in 1513 (Alexander Aphrodisiensis 1513).

protected areas, as several other urban communities did during late Antiquity or the early Middle Ages. The modern citizens are, as a consequence, the direct descendants of those of the Roman period. Their *urbanitas* is pure, since they never mingled with peasants by occupying rural and more secluded areas. The author's emphasis on this point can be tentatively read as an allusion to a 'rival' city such as Capua, whose ancient site had been abandoned in the 9th century in order to found the new city few miles away, on the Volturno river.[39]

Leone is aware that ancient Nola was larger than the modern city, as was easy to conclude by observing the once urban Roman monuments which now lay outside the modern city walls. The entire modern city is located within the ancient urban perimeter. Thus, in chapter 2, Leone draws on his knowledge of mathematics and geometry to calculate—on the basis of his conjectures on the Roman urban perimeter reported in book 1—the difference between the extent of the area covered by the ancient city and that of the modern urban area. Leone concludes that modern Nola is more than six times smaller in extent than its ancient predecessor. Pictures of the urban surface area of both the ancient and modern city and the geometrical representation of Leone's calculations are shown in Mocetto's engraving entitled *Figura praesentis urbis Nolae* [fig. 3].

Chapters 3 to 6 are devoted to modern Nolan territory. They contain: a general description of the territory (ch. 3); a description of the lowlands portion of the territory, with a detailed excursus on how Alfonso I of Aragon used this area to practise falconry (ch. 4); a description of the mountainous parts of the same territory (ch. 5); a description of Castel Cicala and other minor centres of the region, usually called *casali* (ch. 6). In these chapters, there are numerous remarks on the fertility of the soil and on geo-morphological aspects of the territory.

The description of the city moves from the external to the internal areas, so ch. 7 contains a very detailed description of the city walls, in which the names of all the city gates and of the main towers are mentioned; each element (walls, *pomoerium*, towers, etc.) and the distance between each of the elements is systematically measured in feet. With the same precision, in chapter 8 Leone describes the city fortress (*arx*), placed on the southern side of the walls [fig. 36]. After these two chapters devoted to the city's military architecture, chapter 9 contains a description, with some historical remarks, of the Orsini palace

---

39   In the *De Nola*, hints of rivalry between Nola and Capua appear to be found in Leone's account of the construction of the Nolan amphitheatre. See also de Divitiis and Lenzo in this volume, 49–51.

(*regia*) [fig. 28, 29].[40] In this chapter, the baronial role played in Nola by the Orsini family is once again taken for granted and not explicitly mentioned; a clear statement on the subject is found only in book 3, ch. 3, as will be seen. Ch. 10 contains an analysis of the urban streets and a description of one of the most typical public buildings of southern Italian towns, namely the *seggio*, or *sedile* [fig. 33], a portico used for councils and other meetings of the most prominent families of the city,[41] even though the chapter does not mention the function of the building or the families who gathered inside it. Such an omission must be *pour cause*: the explanation probably lies in Leone's idiosyncratic bias against the 'established' nobility of the city, as we shall see below.

Chapters 11 to 14 are devoted to the religious buildings in the city. In chapter 11, in an accomplished rhetorical performance Leone writes a refined description of the Nolan cathedral [fig. 33] behind which the intention to compose a virtuosic architectural *ekphrasis* is easily seen. His description is also a valuable testimony of the Gothic cathedral commissioned by the Count of Nola Nicola Orsini at the end of the 14th century. The cathedral was later re-built in a completely different style. Chapter 12 is a long discussion of the earlier cathedral as it was at the time of St Paulinus. Leone quotes extensively from the letter written by Paulinus (epistle 32) to Severus Sulpicius, in which we find a detailed *ekphrasis* of the four cathedrals in Cimitile. It is noteworthy that Leone misunderstands the epistle, interpreting Paulinus's description to relate entirely to the cathedral in Nola, rather than those in Cimitile. This fundamental mistake gives rise to a long sequence of errors as Leone attempts to superimpose Paulinus's evidence on what he knew about the earlier phases of the cathedral of Nola. He had been probably misled by the fact that Paulinus refers to the church in Cimitile as the *Nolana basilica* (*ep.* 32.17).

The last chapters also deal principally with architectural matters. In chapter 13 Leone lists and briefly describes the remaining churches in Nola; in chapter 14 he explains why several ancient churches are situated below ground level, pointing out how ruins and other factors had raised the city's ground level over the centuries; in chapter 15 we find a noteworthy and detailed analysis of the typical Nolan private house;[42] in chapter 16 Leone lists the benefactors (*benemeriti*) thanks to whom, in his judgement, the main urban and architectural contributions to Nola were accomplished, namely the emperor Tiberius,

---

40 For a discussion of Leone's description of the palace see de Divitiis 2015 and also de Divitiis in this volume, 52.

41 See Lenzo 2014, 178–179; 200.

42 See de Divitiis in this volume. The whole of chapter 15 is transcribed below, in Appendix 3.

who built the temple of Augustus,[43] followed by Nicola, Raimondo, Orso, Niccolò, and Gentile from the Orsini family, as well as a few other benefactors; in chapter 17 Leone inserts a detailed description of all the markets which took place in Nola.

While, as we have seen, the first book of the *De Nola* reflected Leone's antiquarian and scientific education, the second book displays his further expertise, namely his profound familiarity with engineering and architectural matters, a further element which could definitely not have been foreseen from the book title or from his preface.

2.3   *Liber tertius* (ff. xxxx *recto* to lviii *verso*)

In so far as it deals with the modern inhabitants of Nola and the city's politics and society, the third and last book completes the description of *Nola praesens* begun by Leone in book 2. If the second book is focused on geographical, urban and architectural features, book 3 is on Nolan citizens, seen in their public and private life.

Ch. 1 is about the clergy and religious life in Nola; ch. 2 describes funerary chapels and rituals; ch. 3—the most extended in the whole work—contains an impressively long list of the city's most prominent families (*familiae egregiae*), including the names and activities of several members from each of them. As a sort of preface to this chapter, Leone inserts a discussion on the criteria he followed in his selection of the families which merited inclusion in his survey. Leone declares that he considers all those Nolan families *egregiae* who are not involved in manual or servile occupations but, on the contrary, are dedicated to commercial and military activities, or who are professionals with a literary or humanistic education (physicians, lawyers, judges, etc.) (f. xxxxiiii *recto*). Consistent with arguments he had used years before in his *De nobilitate rerum*, here Leone stresses the importance of the *artes liberales* in defining the concept of nobility.[44] He adds, furthermore, that he will also mention some individuals who came from obscure families, but who were nevertheless extremely virtuous. As for the order in which he discusses the families, Leone adopts a 'topographical' system, based on the four districts (*regiones*) into which the urban area was divided (*regio Portellana, Vicanciana, Cortefellana* and *Samuelitana*), mentioning the families on the basis of the *regio* they live in. His discussion of the Orsini family seems to reflect a tension between understatement and

---

43   See Suetonius, *Tiberius* 40.

44   Leone 1525. The *De nobilitate rerum* was written before the *De Nola*, but published only after the author's death, in 1525: see also below. For a full discussion of Leone's chapter on Nolan families see Vitale in this volume.

encomium: on the one hand, Leone lists and discusses this family among oth-ers while describing the Portellana neighbourhood, where the Orsini palace was located; on the other hand, in the same passage we find the only explicit reference in the whole of the *De Nola* to the fact that the Orsini were the bar-ons who ruled the city; here Leone also inserts some short courtly *encomia* of individual family members.

After a chapter on the presence of the Jews in Nola and their expulsion (ch. 4), Leone focuses on the citizens' main *studia* or 'occupations' (ch. 5). Leone opens this entirely encomiastic chapter by claiming that the Nolans of his time still preserved the elegance and refined style of life which had char-acterised them in antiquity (f. 1 *verso*), with reference to the parallel chapter inserted in book 1, and discussed above, in which he dealt with the ancient Nolans' *elegantia*.[45] Leone's intention of underlining the continuity from Roman to modern times could not be clearer. As for the *elogium*, he inserts such statements as "Nola produces more beautiful men and women than any other of the surrounding places" ("pulchriores et viros et mulieres Nola gignit quam quaevis ex circumpositis locis"), or "The Nolans are ignorant of faction, homicide and treachery" ("factiones quoque nesciunt Nolani, neque homicid-ia nec proditiones", or "Incredibly, prostitutes from Nola have never been seen anywhere" ("illud quod mirum est nusquam visas esse Nolanas meretrices"), and so forth.[46]

The last part of this chapter is a sort of gallery of famous humanists who, even if they were not Nolans, appreciated Nola and spent some time there. The list includes: Giovanni Pontano, Pietro Gravina, Antonio de Ferrariis, Lorenzo Valla, Gianfrancesco Caracciolo and others.

A central part of the book (chapters 7 to 10) concerns political and eco-nomic matters, and gives a remarkable description of the whole institutional structure of the city. For the first time in the work, readers are given an expla-nation of the Count's role in civic life, and also the part played by the nobility, although Leone avoids listing the narrow number of families who actually con-stituted this class. In explaining some aspects of political life in the city, Leone is on occasion 'obliged' to confront the fact that some officials (*magistratus*) were chosen from the nobility (*ex primariis Nolanis*), despite the fact that this nobility does not explicitly coincide with (although it is certainly included in) the number of notable members of Nolan society (*egregii*) listed in chapter 3. Such an 'overlooking' of this political and social dimension can probably be attributed to a deliberate purpose on Leone's part to understate the role played

---

45   Already in book 1, however, many remarks focus on modern times.
46   All quotations are drawn from f. 1 *verso*.

by these families, an intention which springs from the wider and more liberal idea of nobility which emerges from chapter three in the third book.

Chapter 7 focuses on local officials (*magistratus*), whose tasks and characteristics are described in detail; it is also specified if each *magistratus* is nominated by the Count—as was more frequent—or by the citizens. Chapter 8 is about the local taxes. Chapter 9 deals with marriages and dowries. In chapter 10 Leone describes the theatrical performances and jousts which took place in the city, both of which are presented by Leone as direct descendants of the Roman spectacles held in theatres and amphitheatres.

Chapters 11 to 14 are devoted to Nolan folklore:[47] in chapter 11 the author describes the procession made by all the Nolan priests on 27 April, which ended in the cathedral with a ritual *salutatio* to the bishop, a tradition that Leone's classicism connects with the ancient pagan rituals which were once celebrated in the temple of Jupiter and were subsequently transformed into a Christian ceremony. In chapter 12 a ritual struggle among peasants which took place inside the cathedral is described, consisting in two teams which competed to take possession of a pig. Leone interprets also this ceremony as an ancient pagan ritual connected with the cult of Jupiter, resembling, in Leone's opinion, a gigantomachy, which survived in Christian form. In so claiming, the Nolan humanist explicitly contradicts Giovanni Pontano, who on the contrary had branded a similar ritual fight celebrated in the cathedral of Naples during his time as a bestial affair.[48] Chapter 13 describes a folkloric battle fought once a year by adolescents, which often degenerated into a sort of communal riot, without losing, nevertheless, its ritual character. Chapter 14 gives a careful description of the folkloric ritual of the grape harvest, focusing in particular on the obscene language used by the peasants (*vindemiatores*) while working, and the obscene insults they hurled against every kind of authority figure, whether political or ecclesiastical. Thanks to its ritual context, this behaviour, reinforced by the quantities of wine they drank, was tolerated by everybody. Predictably Leone traces the origins of this feast back to the ancient ceremonies performed in Bacchus' honour.

Leone's approach to the Nolan society of his own time also reveals original aspects of his prose style, in particular a sort of 'ethnographic' tendency which enables him to draw an especially vivid portrait of Nola's institutions and ceremonies. In doing so, his writing seems to have assimilated the lessons of classical historiography in a highly refined and subtle way.

---

47   Full discussion in Imbriani in this volume.
48   Here Leone refers to a passage from Pontanus's dialogue *Charon*: Pontano (ed. Privitera) 1943, 25–26.

3   The Genre of the De Nola: Between Antiquarianism, Chorography, and Encomium

The simultaneous presence in the same book of different methods and approaches should lead us to explore in more detail the genre to which the *De Nola* belongs and the models which lay behind it. We have seen how in the *Praefatio* Leone is eager to present his work as a purely descriptive (especially historical) treatise, a purpose which is revealed by the title with its *de* + ablative formula. This 'labelling' seems however to creates more problems than it solves, since at least three 'spirits' animate Leone's study of the city, meaning that, as a consequence, his work appears to be placed at the intersection of three codified genres, each of which was highly successful when Leone was writing: antiquarian literature, chorography, and rhetorical works in praise of cities (*laudationes urbium*). We will briefly discuss the influence of these genres on the *De Nola* on the assumption that each of them was regarded by Leone as a fundamental but not constrictive model.

To read the classical past of a city through its extant ruins and with the help of classical sources—in other words, to take an antiquarian point of view—constitutes an approach which, when Leone published his *De Nola*, had an indisputable precedent in the antiquarian works of Biondo Flavio of Forlì.[49] An 'archaeological' approach to Rome had already been attempted (after Petrarch's famous and pioneering epistle *Fam.* 6.2 to Giovanni Colonna)[50] by Poggio Bracciolini in his *Ruinarum urbis Romae descriptio*,[51] but Biondo's detailed reconstruction and celebration of Rome in the *Roma instaurata* and, later, in the *Roma triumphans*,[52] along with the entry *Rhoma* in Tortelli's

---

49   A biographical profile in Fubini 1968; see also the essays collected in the recent volume Mazzocco, Laureys 2015. A complete edition of his works is planned by the Italian *Edizione nazionale delle opere di Biondo Flavio*, in which several volumes have already been published. Several of Biondo's works are also planned for publication as part of the *I Tatti Renaissance Library* Series. A useful and up-to-date bibliographical tool is www.repertoriumblondianum.org.

50   On Petrarch's antiquarianism see Mazzocco 1977.

51   Written between 1424 and 1431, and inserted in his *De varietate fortunae* in 1448, the *Ruinarum urbis Romae descriptio* was printed for the first time in 1513 (Bracciolini 1513, f. 50 *recto*–52 *recto*). A modern edition in Bracciolini (ed. Boriaud) 1999.

52   Written in 1446 and first printed in 1471 (Biondo 1471), the *Roma instaurata* was a fundamental book for the whole of the early modern period. A modern edition and French translation in Biondo (ed. Raffarin-Dupuis) 2005–2012, but on the need for a new edition see Della Schiava, Laureys 2013. The *Roma triumphans* was completed in 1459 and first

De orthographia,[53] had changed forever the way of describing urban antiquities.[54] These works suddenly rendered obsolete all the previous medieval texts on the *mirabilia* of Rome, a genre which had enjoyed enormous success over several centuries, functioning as a kind of 'pilgrim's guide' to the Eternal City and focused essentially on the Christian attractions of Rome.[55] Biondo's influence on the *De Nola* seems significant above all because his celebration of Rome's antiquity does not obliterate the Christian stage of the city's urban development, though the magnificence of the ancient monuments is seen as incomparably superior to any later building. His view of the present is positive and characterized by a greater interest in the continuities rather than the discontinuities with its classical past, exactly as Leone describes and praises both ancient and modern Nola at the same time. When we consider the structure of the *Roma instaurata*, and its arrangement into books and chapters, several similarities with Leone's work are detectable.

A certain degree of influence on Leone was certainly exercised also by later antiquarians:[56] Pomponio Leto, for instance, whose studies of Rome's chronology and whose collections of antiquities furnished a model for several humanists;[57] or Giovanni Pontano, leader of an academic circle (the so-called *Pontaniana porticus*),[58] in which the antiquities of Naples were accurately discussed, and great attention was paid to antiquarian issues and mythographical sources connected with the origins of cities—we think, for instance, of the description of Naples that Pontano inserted at the end of his *De bello Neapolitano*.[59]

---

        printed in 1473 (Biondo 1473). For both these works new editions are in progress: see note 49.

53   Tortelli 1471a e 1471b, s.v. "Rhoma". Though a critical modern edition of the entire work is still lacking, Tortelli's entries are indexed in Donati 2006, and the entry "Rhoma" has been edited in Tortelli (ed. Capoduro) 1999.

54   Not to mention, of course, Leon Battista Alberti's *Descriptio urbis Romae*, which offered a meticulous topographic reconstruction of the ancient city (edition in Alberti [ed. Boriaud, Furlan] 2005). On Alberti's influence on Leone see Lenzo in this volume.

55   On this topic, with editions of some of these works, see Miglio 1999; Accame, Dell'Oro 2004; Barkan 2010, 845; Campanelli 2011; see also the essays collected in Hamilton, Riccioni 2012.

56   On the Renaissance literature focusing on the antiquities of the Italian centres Weiss 1969 is still useful as a general overview.

57   Biography and bibliography in Accame 2015.

58   Biographical profile in Figliuolo 2015.

59   Pontano (ed. Summonte) 1509, ff. [G v] *recto*—[G vii] *verso*.

Pontano was undoubtedly a relevant model for Leone, who cites him twice in the *De Nola*, thus showing that he was familiar with this humanist's works.[60]

Probably another antiquarian work was also influential on the structure and other formal aspects of the *De Nola*, namely Francesco Albertini's *Opusculum de mirabilibus novae et veteris urbis Romae*, printed a few years before the *De Nola*, in 1510.[61] Albertini describes, albeit briefly, the major antiquities of Rome, arranging the materials in such a way that may constitute a parallel to Leone's work, though only for books 1–2.[62] Albertini's influence, furthermore, is perhaps deducible also from the use of the word *opusculum* in the title of the *De Nola*.

Alongside the authors and the works mentioned above, one must not underestimate the influence of the antiquarian interests which developed in Venice after Pier Paolo Vergerio wrote his *De Republica Veneta*[63] in 1404 and, later, Biondo Flavio his *De origine et gestis Venetorum*.[64] Between the end of the fifteenth and the beginning of the sixteenth century, this local antiquarian interest was still considerable, as is shown by works like Bernardo Giustinian's, *De origine urbis Venetiarum rebusque eius ab ipsa ad quadringentesimum usque annum gestis historia*,[65] or Marin Sanudo's *De origine, situ et magistratibus urbis Venetae*,[66] and, above all, with the printing in 1502 of Marco Antonio Sabellico's *De situ Venetae urbis libri tres*, written more than a decade before.[67]

Biondo Flavio is a detectable model also for the second genre which deeply influenced the *De Nola*, i.e. chorographic literature. Though focused on the description of the entire Italian peninsula rather than a single city, Biondo's *Italia illustrata* mixed the use of classical sources with the author's own examinations or, as an alternative, the testimonies of local residents, and provides a

---

60 Leone 1514, bk. III, ch. 6, f. li *recto*; bk. III, ch. 12, f. lvi *verso*.

61 Albertini 1510. On this book and its author see Bianca 2011; Amato 2012.

62 A certain influence on the *De Nola* (also perceptible in the book's layout and on the title page) may perhaps have been exercised by the *Notitia regionum urbis Romae* edited by Giano Parrasio in Milan in 1503 and attributed by him to a fictitious Publius Victor (*Publius Victor* [ed. Parrasio] 1503), a book which enjoyed an enduring fame in the sixteenth century. I am thankful to Adolfo Tura for suggesting this parallel to me.

63 Ed. in Robey, Law 1975, 38–50.

64 Written in 1454 (Fubini 1968), and first printed in Biondo 1481–1482.

65 Printed after the author's death (Giustinian 1492), written about in 1477–1481 (see Pistilli 2001).

66 Edited in Sanudo (ed. Caracciolo Aricò) 2011.

67 Sabellico 1494; on this work see Modesti 2009, with bibliography.

short account of each place describing their past and modern features.[68] Thus, antiquarian method is balanced by an actual exploration of the territory. When Leone wrote his *De Nola*, Biondo's chorographic method had already become predominant, and had already had such noteworthy epigones as, among others, Pietro Ranzano and Raphael of Volterra, although the method of the latter was more 'bookish', and the work of the former remained unpublished.[69] All these works, however, described extended territories and not single towns: in such a systematic work as the *Italia illustrata*, discussions of specific cases are necessarily cursory, as we also find in Ranzano's or Maffei's works. As is well-known, furthermore, Biondo's survey is not complete, since he was not able to finish his description of the larger part of southern Italy. Chorographic works operating on a less extended scale were however also available: an author who exercised a certain influence on Leone was Antonio de Ferraris of Galatone, hence called Galateo, an Apulian humanist and physician who befriended Leone in Naples and probably introduced him into Pontano's circle.[70] Slightly older than Leone—and mentioned as a close friend by Leone himself in *De Nola*[71]—, Galateo was well known in Venice, as witnessed by the fact that he himself wrote an encomium of Venice,[72] dedicated to the Venetian politician Luigi Loredan, and that Ermolao Barbaro dedicated to him his translation of Themistius' paraphrases to Aristotle's *Physics*.[73] Between 1506 and 1511 Galateo wrote a remarkable description of his homeland, the Salento region, corresponding to the ancient territory of Iapygia. In this work, the *De situ Iapygiae*, Galateo describes (and also praises) the main centres of the region by

---

68　　Biondo's *Italia illustrata* has been recently edited and translated into English: Biondo (ed. Castner) 2005–2010; Biondo (ed. White) 2005–2016. A new critical edition, with ample commentary, is in Biondo (ed. Pontari) 2011–2017 (the volume including the section on Campania is still in progress).

69　　Maffei 1506; for Pietro Ranzano, see the modern edition Ranzano (ed. Di Lorenzo, Figliuolo, Pontari) 2007.

70　　On Galateo see Romano 1987. On his friendship with Leone, see Sica 1983, 13–14; Defilippis 1991; Vecce 2000.

71　　Leone 1514, bk. III, ch. 6, f. li *recto*: "Antonius quoque Galatheus, utroque literarum charactere ac scientia omni vir maxime circumseptus, Nolam libentissime veniebat semperque domum nostram divertens. Is praeter caetera dicere solebat Nolam sibi solam Syracosias esse. Nam Syracosiae videntur appellatae quod vel invitos homines ob amoenitatem trahant". Galateo is also the protagonist in one of the *problemata* edited by Leone in his *Opus quaestionum* (Leone 1523, probl. 306).

72　　An edition in Tateo 2006, 320–323.

73　　Barbaro 1481, f. a1 *recto–verso*. On Galateo's connection with the Venetian milieu see Defilippis 2006 and Nuovo 2006, both with ample bibliographical references.

focusing on their Greek or even pre-Greek origins, describing ancient monuments, reading ancient Roman inscriptions, and so on.[74] Since Galateo was a physician too, like Leone, the *De situ Iapygiae* is also full of scientific notes on plants, animals, diseases, physical phenomena, and so forth. It is possible that Galateo's work provided Leone with a useful model, since it offered an accomplished and cultivated chorographic and antiquarian description of a fairly limited area, suitable for treatment in a monograph on the subject. Since the *De situ Iapygiae* was not printed before 1557, however, a direct reading of the work in manuscript by Leone is possible but conjectural. It is less problematic to suggest that there was a more general and reciprocal influence between the two humanists, who frequented the same milieu for a long time and had several interests in common. In 1513, furthermore, while Leone was attending to the publication of the *De Nola*, Galateo completed the composition of his *Callipolis descriptio*, a short (encomiastic) description of the Salentine city of Gallipoli.[75] Thus the idea cannot be excluded that the two humanists influenced each other in writing and reflecting about the description of cities.

The question of the encomiastic purpose of urban descriptions like those by Galateo introduces us to the third relevant model for the *De Nola*, the so-called *laudationes* or *laudes urbium*, a genre which often overlaps with the *descriptiones urbium*, the distinction between these two categories being unclear and undefined.[76] We have seen in analysing the preface of the *De Nola* that Leone sought to distinguish his work from the rhetorical genre of the *laudatio*. When he came to write his patriotic account of his native city, Renaissance literature had already produced several masterpieces in this genre, the most famous being Leonardo Bruni's *Laudatio Florentinae urbis*,[77] to which Pier Candido Decembrio's work on Milan *De laudibus Mediolanensium urbis panegyricus* was a response.[78] These humanists re-interpreted a genre that in the Middle Ages had flourished, attaining its apogee with works like Bonvesin de la Riva's *De magnalibus urbis Mediolani* and others.[79]

Unlike the medieval *descriptiones*, which generally devoted their accounts to a real (though summary) urban description, with carefully compiled lists of e.g. churches, noble families, and so on, the humanistic type of *laudatio*

---

74 A modern annotated edition translated into Italian in Galateo (ed. Defilippis) 2005.

75 Edited in Paone 1974; Zacchino 1977.

76 On this subject see Classen 1980.

77 Written in 1402–1404, modern edition in Baldassarri 2000.

78 Written in about 1435–36, ed. in Petraglione 1907.

79 On this genre see Hyde 1965–1966; Fasoli 1973; Classen 1980. An edition of Bonvesin de la Riva's work on Milan in Bonvesin de la Riva (ed. Chiesa) 2009.

urbis marked a return to classical models, by referring to more ideal features, such as the moral virtues of the citizens, the city's classical origins, its political constitution, and so forth. The humanists benefited from a large number of ancient models: praise of a city was an ancient oratorical genre of which there are both Greek and Latin examples.[80] As for the contents of this type of *laudatio*, ancient encomia did not necessarily include a systematic urban description: works like Isocrates' *Panegyricus* and *Panathenaicus* are above all praises of the moral and political virtues of the citizens, and are very selective in mentioning monuments and buildings. When in the third century CE Menander Rhetor summarised the rules of the genre, he does not mention the need for a detailed description;[81] the assumption is perhaps that this was more a matter for historiographical works. It is not by chance that we find a (quite encomiastic) description of Rome not in such a work as Aelius Aristides' *To Rome*,[82] but in Strabo's chapter devoted to the *aeterna Urbs*, i.e. in a geographical composition.[83]

Bruni's *laudatio* of Florence, which is modelled above all on Aelius Aristides' *Panathenaicus*,[84] contains no systematic urban description. If we consider the whole question from this point of view, Leone's desire to distinguish his work from the *laudatio* genre is not without its reasons, but if, on the other hand, we take into account all the encomiastic valuations of the exceptionally virtuous character of Nolan women and men which the book contains (like those mentioned above), then we can appreciate how deeply this genre actually influenced Leone's work. It should however be noted that a detailed comparative analysis of the *De Nola* and explicitly encomiastic Renaissance works like those of Bruni or Decembrio is still lacking, though it would be a promising path of study.

4   The *De Nola* as a Humanistic Work: Leone's Use of Greek and Latin Sources

In his elaboration of the structure of his book, Leone drew on a large number of classical sources. What follows is a brief analysis of his explicit citations of

---

80   Pernot 1993, 178–215.

81   Menander Rhetor 1.346.26–367.8 (ed. Russell, Wilson 1981, 33–74).

82   Aristides or. 36 K.

83   Strabo, 5.3.8.

84   Bernard-Pradelle 2000; Revest 2007; Fontanella 2013, who also detects in it an influence of Aristides' speech *To Rome*.

Latin and Greek texts, leaving aside for the moment his implicit allusions or 'hidden' references, which characterize his prose style as well—as seen above in the allusion to Sallust contained in the *Praefatio*.

Studying the working methods of Renaissance Latin writers means above all dealing with the problem of what type and what format of literary sources they consulted while engaged in composition. From this perspective, Leone is apparently a son of the so-called Gutenberg Revolution, since most of the sources he quotes are drawn from printed editions. This is true above all for Greek quotations, i.e. for those sources with which Leone had become familiar only shortly before writing the *De Nola*, since he learned Greek after his arrival in Venice. If the reliance on printed books was a customary practice among humanists in the second decade of the sixteenth century, Leone's case is a particular one: his close friendship with Aldus Manutius and numerous other *Philellenes* allowed him to consult several printed *Aldinae*, and presumably to follow (and benefit from) editorial works which were still in progress. More generally, Leone's participation in such an intellectual milieu as Aldus' circle put him in the best position to take advantage from an impressive number of printed editions and from the expertise and competence of their editors.

More complex is the case of the (especially Latin) works cited by Leone in the *De Nola* which he undoubtedly knew since the time he lived between Nola and Naples, e.g. the Latin sources about Nola during Roman times, such as Livy or Silius. Future studies may perhaps show if these citations are primarily based on printed editions consulted in Venice, or if they depend on materials circulating in the Neapolitan ambience: in the former case, Leone would have built *ex nihilo*, in Venice, the *corpus* of the sources that he used in composing his *De Nola*; in the latter, he would have drawn on notes made while he was still living in the Kingdom of Naples, which he later re-used in Venice.

The following account offers an outline of the most significant or intriguing sources used by Leone, while a repertory of the passages cited is furnished in the table below.

Predictably, most of the quotations from Latin and Greek authors are located in book 1, in which Leone focuses on Nola's ancient past. A first group of sources includes works that explicitly mention Nola or report episodes from Nolan history. Livy has of course pride of place: here Leone could find such patriotic figures as Lucius Bantius or Erennius Bassus, and in general accurate and detailed accounts of the Campanian phase of the Second Punic War, in which Nola pursued a loyal (and farsighted) policy towards Rome.[85] An important role was also played by other Latin authors who collectively constitute a

---

85   See above, the discussion of book.

kind of 'canonical' encyclopaedia for Roman history and topography: Pliny the Elder, Solinus, Suetonius, etc. Some more extemporary Latin citations are apparently due to Leone's interest in physiology and medicine, as is the case with some citations from Celsus, Columella, and Varro,[86] all dealing with botanical matters, a field closely connected with Leone's knowledge of medical plants.

Leone refers to late antique and/or Christian sources very rarely: an exception is a reference to Paulus Diaconus' *Historia Romana*, in which Leone finds the information that Nola was destroyed by Vandals and "Mauri".[87] A major exception to this rule is represented by Paulinus of Nola, whose poetic letter 32 is extensively reported *verbatim* with most of the rest being paraphrased. Leone could hardly not have taken account of so sophisticated a writer as St. Paulinus—a Gaul by birth and bishop of Nola from 409 to 431: Paulinus' presence in the *De Nola* constitutes a tribute not only to the man who was the city's outstanding religious and literary glory, but also to what was a detailed and accurate local source on the history of the place. Paulinus was the first to compose a description of the churches of Cimitile, although Leone was wrong in taking it to be a description of the cathedral in Nola, as we have seen above (see the discussion of book 2 ch.12).

Leone's use of Greek sources can serve to highlight his humanistic working method. In general, the presence of the Greek language is widespread throughout the whole work: Leone often refers to single Greek terms, written in Greek characters, in order to explain the origins of words, denominations, and toponyms. The case of more extended passages drawn from Greek authors is different: when not merely paraphrased, Greek sources are generally quoted by Leone through Latin translations. The only exceptions to this are one quotation from Dioscorides and one from Libanius, both of which will be discussed below.

---

86   Celsus, quoted in bk. I 2, f. vii *recto*, without a precise textual reference, on the *palmula*, mentioned by the ancient physiologist in several passages of his *De medicina* (see 1.2.9; 2.20.1; 2.24.2, etc.); Columella, *De re rustica* 10.133: bk. I, ch. 11, f. xvi *recto*; Varro, *De lingua latina* 5.31: bk. III, ch. 14, f. lviii *recto*.

87   Paulus Diaconus, *Historia Langobardorum* 15.17, quoted at bk. I, ch. 6, f. x *verso*. The passage is important also because a few lines below, at 15.18, Paulus gives an account of Paulinus of Nola's (legendary) period of captivity in Africa. This means that the whole section was a relevant source for the history of Nola, and this is probably the reason Leone was familiar with it. The *Historia Romana* had already been published in several editions, from 1471 onwards. It is however noteworthy that Leone's friend Giovan Battista Egnazio in 1514 was preparing an edition which would be printed by Aldus' heirs in 1516 (Egnazio 1516b).

If compared to the corpus of cited Latin authors, the Greek corpus used by Leone is perhaps less homogeneous. We find 'predictable' sources like the geographers Strabo or Stephanus of Byzantium, and some citations from Greek medical authors may also be considered unsurprising, given Leone's professional education. Perhaps unexpectedly, Galen is quoted only once, but this citation involves a remarkable extended translation of a passage dealing with ancient Stabiae from the *Methodus medendi* 5.12: here Leone's effort is all the more noteworthy as it is elaborated directly on the basis of the Greek text (probably taken from the *editio princeps* of 1500), since Latin translations of this Greek work are not attested for this period.[88]

Other citations or allusions are more unexpected and were probably influenced by the intense editorial activity developed in the Venetian publishing milieu under the auspices of Aldus Manutius. As seen above, the joint quotation of Isocrates and Aelius Aristides in the *Praefatio* can be traced back to the 1513 *editio Aldina* of the two Greek orators. An analogous influence can also be conjectured to explain the quotation from the Greek text of Libanius' *Argumenta Demosthenis*: why did Leone refer to such a particular text in order to demonstrate the Athenian origins of the Euboeans? The answer probably lies in the fact that in 1513 the *Argumenta* were re-printed by Aldus together with Demosthenes' speeches,[89] meaning that the works of Demosthenes and Libanius must have been circulating among the *Philellenes* in Aldus's circle in those very same years.

Again, the use of the Greek text for the citation of a passage from Dioscorides, in which the ancient physician notes that the *melilotus* plant is also found near Nola,[90] is certainly based on one among the numerous editions and translations of this author which had been available since 1499, when Aldus printed the Greek text. One may add that Leone's closest friend in Venice, Giovan Battista Egnazio, was preparing in this period a work of *annotamenta* on Dioscorides which included the fifteenth-century Latin translation

---

88 Leone 1514, bk. I, ch. 11, f. xv *verso*. I have not found Latin translations of Galen's *Methodus* before 1519 in searching the main research resource on this subject, namely During's 'census' of Renaissance editions and translations of Galen (During 1961). The Cretan humanist and printer Zacharias Calliergi was responsible for the *editio princeps* of Galen's *Methodus*, published in Venice (Galen 1500). In general, Leone's use of Galen and the translation of this passage in particular deserve further investigation.

89 *Demosthenis Orationes* 1513. This is an exact reproduction of a previous edition, published by Aldus in 1504; see Renouard 1834, 47–48.

90 Dioscorides 3.40.1. The passage quoted by Leone: μελίλωτος· κράτιστος ὁ ἐν Ἀττικῇ καὶ Κυζίκῳ καὶ Χαλκηδόνι γεννώμενος, κροκίζων καὶ εὐώδης· φύεται δὲ καὶ ἐν Καμπανίᾳ περὶ Νῶλαν τηλίζων καὶ ἄτονος κατὰ τὴν εὐωδίαν.

by Ermolao Barbaro. This book was published in 1516, but it is highly probable that Leone followed his friend's editorial work over the years, especially given his own interest in medicinal plants.[91]

Leone's use of Dio Cassius is also very interesting. He includes in chapter 2 of book 1 a Latin translation of an extended passage drawn from Dio's book 66, namely the account of the famous eruption of Vesuvius in 79 CE. Leone could not have read a Latin translation of book 66 in its entirety since one did not exist (it was transmitted only in Byzantine *epitomae*), but Giorgio Merula had already translated a few decades earlier the passage relating to the eruption. This extract was included in the 1503 edition of Merula's translation of Dio's books concerning Nerva, Trajan, and Hadrian with the title *Conflagratio Vaesevi montis ex Dione Merula interprete*; In 1510, the *Conflagratio* was reprinted in Venice, at the end of the edition of Biondo Flavio's works.[92] It is clear that Leone's use of Dio Cassius in this case cannot be understood without reference to the influence of Merula's translation of the passage, probably *via* the reprint of 1510, although only further investigation could reveal the precise relationship between the two translations. Even more intriguing are Leone's references to Dio 56.29 and 56.31, two short passages—amounting to no more than a few lines—concerning the death of the Emperor Augustus at Nola, in which Leone's Latin translation depends directly on the original Greek text.[93] In 1514 book 56 had neither been translated into Latin nor had the Greek original been published, meaning that Leone must have drawn these pieces of information from another source, possibly even from a Greek manuscript of the work, perhaps inspired by commentaries to Suetonius' *Lives*, like those by Marco Antonio Sabellico and Filippo Beroaldo.[94]

Other citations appear to recall Leone's recent study of Greek, and sound like a sort of homage to Marco Musuro, who was Leone's teacher in the language.[95] This is probably the case for a quite unexpected citation of Synesius' epistle

---

91   Egnazio 1516a.

92   Censorinus, Dio *et alii* 1503; Biondo 1510 (Merula's translation is at ff. CXXXXV *verso*– CXXXXVI *verso*).

93   Miletti 2016b.

94   On Suetonius' commentaries in the Renaissance period see Pellegrini 2007–2008. I wish to thank the anonymous referee for suggesting to me this hypothesis, and Marijke Crab for discussing with me this problem, also by pointing to me (on the basis of Fera 1988, 85– 87) that Battista Guarini was the first humanist to apply a systematic reading of Suetonius and Dio in parallel.

95   A new and detailed account of Musuro and his philological activity in Speranzi 2013 and Ferreri 2014. For Leone's friendship with Musuro see the introduction to this volume.

1.1–3, drawn from the *editio princeps* of several authors of epistles, edited by Musuro in 1499 and also printed by Aldus.[96] Leone's reference to the plot of Sophocles' *Ajax* (3.12, p. lvi *verso*) might be connected with the *editio princeps* of Sophocles' plays printed by Aldus in 1502, but could also be a reference to the courses on the Greek tragic poets and on Sophocles alone held by Musuro respectively at Padua in 1506/1507 and in Venice over a year in the period between 1512–1516.[97]

The evolution of Leone's philological and antiquarian mastery of his classical sources is detectable through a comparative reading of the *De Nola* with Leone's *De nobilitate rerum*, a work which was published after his death, in 1525, but the composition of which can be dated presumably to the end of the fifteenth century (about fifteen years before the *De Nola*), since it was clearly written for a Nolan (or Neapolitan) audience. This work was published by Leone's son Camillo, who claims, in the preface, that he decided to print the work after it had remained among his father's unpublished manuscripts for a long time. A modern edition and/or modern study of this interesting work is still lacking, but if we read it in parallel with the *De Nola*, we find a very different use of classical sources. In particular, from the almost complete absence of direct Greek quotations we understand that Leone's passion for Greek studies was still in its infancy. The main models are Latin and can be identified with such 'conventional' sources as Cicero, Pliny the Elder, etc., besides a general Latin-Aristotelian philosophical knowledge. In the *De Nola*, as we have seen, the variety of the sources Leone uses is remarkable, while a major role is played by Greek classical authors whom Leone read without having to resort to the medium of Latin translations.

Leone's approach to Greek and Latin sources also influenced the prose style of the *De Nola*, which is not characterised by a complete humanistic adherence to pure classical Latin prose but is rather a language which, while certainly influenced by classical authors, also reveals Leone's reading of medieval treatises of medicine and philosophy as well. Leone's style is however sober and direct, principally shaped by historiographical and scientific models, though it can easily turn to more lyrical and melancholic tones when describing sentimental matters, like the death of Beatrice de Notariis, or to a more emphatic and

---

96   *Epistolae* 1499. This edition is divided into two tomes, the first of which contains Synesius' epistles. An analysis in Ferreri 2014, 112–131.

97   Sophocles 1502, an edition often attributed to Marco Musuro but to which in fact it is probable that Musuro never contributed: see Ferreri 2014, 395–401. On the courses at Padua (tragic poetry) and Venice (Sophocles) see Ferreri 2014, 437–438; 451–452.

rhetorically ornate style when praising his hometown or in the pages of the *Praefatio.*

5   Conclusions

The analysis of Leone's use of classical sources in the *De Nola*, especially when considered in the light of the work's purpose and structure, enables us to draw some conclusions about his method and his personality as humanist and antiquarian. The *De Nola* is, on the one hand, undoubtedly characterised (and often limited) by a patriotic intention which sometimes leads Leone to misunderstand, or over-interpret, the data he deals with, but, on the other hand, it also comes across as a 'multidisciplinary', encyclopaedic, stratified and complex work. When the book was written, Leone was a mature doctor and humanist who had begun a 'second life' in Venice, after his 'first life' between Nola and Naples. The stratification of such a complex education is readable, between the lines, throughout the work. A first layer is constituted by a scholastic-Aristotelian education, which merges with the Latin humanistic readings made under Pontano's influence. A second layer is the fruit of Leone's Venetian philological and antiquarian education, on which the influence is clearly perceptible of his Hellenic and philhellenic teachers and friends, above all Marco Musuro, Giovan Battista Egnazio, Girolamo Aleandro and Aldus Manutius. And it is this Venetian context which can be said to have had most influence on Leone's book. His extensive use of Aldine editions and the wide-ranging collaboration with several Venetian intellectuals can be considered as aspects of Leone's participation in one of the most culturally advanced and pioneering milieus in Renaissance Europe in the second decade of the 16th century.[98]

If we compare this work to the extant antiquarian and chorographic works dealing with descriptions of cities which were available in 1514 we can better appreciate Leone's innovative approach. Although his humanistic method of working may be criticised as lacking a comprehensive philological grounding, it is nevertheless characterized by a remarkable ability to combine scientific and physiological questions arising from his Aristotelian education with a sensitivity towards the literary and visual arts and architecture and an erudite reading of classical sources. An originality of approach which can be detected in several aspects of the *De Nola*, among which three in particular deserve to be

---

98   The profound involvement of Leone's work with the refined and pioneering Aldine milieu
         was first stressed by Roberto Weiss (1958, 178–179).

singled out. First, the adoption of a 'scientific' approach to the description of cities and territories: Leone enhances the antiquarian methodology, based on the examination of archaeological evidence and the interpretation of written sources, with the tools provided by his scientific and technical expertise. On several occasions he inserts arguments drawn from botany, geology, medicine and so forth into antiquarian discussions. We have also seen that he often resorts to meticulous geometrical calculations, and in general to his extensive architectural knowledge in order to describe correctly and analyse both ancient and modern buildings.

Second, the detailed comparative focus on both ancient and modern periods, an approach to which Leone explicitly refers in the *Praefatio*,[99] and which is made clearly visible through the use of sophisticated and innovative graphic representations of both ancient and modern Nola.

Third, the constant systematic use of Greek sources, in addition to the more 'canonical' Latin ones: in line with the studies and enquiries he undertook in Venice, Leone handles several Greek editions from which he draws direct or indirect quotations, and, generally, uses Greek sources in order to complete and strengthen his arguments.

The sum of all these features makes the *De Nola* fundamental reading if we wish to understand the many cultural and literary processes unfolding at the heart of the Renaissance. Far from being a work of merely local interest, the *De Nola* is an outstanding example of how, in the early sixteenth century, the description of a country or a city was thought to be fully achievable only by means of an all-encompassing approach which, in order to give an account of the human and material issues involved as well, draws on the methodologies of disciplines connected to different fields of human knowledge, such as philology, rhetoric, architecture and medicine.

---

99   See Appendix 2. In the preface, Leone claims that one of his main purposes was to provide contemporary readers with a resource they could use to understand their past, in the same way that the description of modern Nola would in turn enable future readers to better understand what would become theirs. Similarly at the beginning of his description of the typical house of the city (bk. II, ch. 15) Leone declares that, since no private house from Roman times has survived and the same fate might befall the houses of his own time, he will describe them in detail so that future readers will be able understand how their ancestors lived and in what kinds of house (see Appendix 3).

TABLE 1 *Classical sources quoted in the* De Nola

M= mention of a work, without references to specific passages; R= reference to a specific passage of a work, without direct citation or paraphrasis; P= paraphrasis; GT= citation of the Greek text; LT= citation of the Latin text; LTransl= Latin translation.
Greek sources in bold.

Book 1

Passage or work	Type of citation	Position in the *De Nola* (chapter and folium)	Possible printed or ms. source (only for the Greek works)
Isocrates, *Panathenaicus*	M	*praef.*, ii^r	*Oratores Graeci* 1513
Aelius Aristides, *Panathenaicus*	M	*praef.*, ii^r	*Oratores Graeci* 1513
Dio Cassius, 66.21.1–23.3–4	LTransl	1, iiii^v–v^r	Dio 1503 Also a connection with Merula's translation in Censorinus, Dio *et alii* 1503; repr. in Biondo 1510.
Pliny the Younger, 6.16	R	1, v^r	
Celsus, 1.2.9; 2.20.1; 2.24.2 *et al.* (*palmula*)	R	2, vii^r	
Virgil, *Georg.* 2.217–225	T	2, vii^v	
Gellius, NA, 7.20	P	2, vii^v	
Virgil, *Aen.* 7.740	T	2, viii^r	
Servius, *Aen.* 7.740	T	2, viii^r	
Dioscorides, *De materia medica* 3.40.1	GT and LT	2, viii^v	Dioscorides 1499 or reprints. Also a connection with Egnazio's work on Dioscorides? (Egnazio 1516a)?
Livy, per. 73	T	3, viii^v	
Livy, 37.10	T	3, viii^v	
Pliny the Elder, 3.43	T	3, viii^v	
Strabo, 5.4.7	LTransl	3, viii^v	
Silius, 12.161–164	T	3, viii^v	
Trogus apud Just., 20.1.13	T	3, viii^v–ix^r	
Stephanus Byzantinus, s.v. Χάλκις, p. 683–685 Meineke	P	3, ix^r	Stephanus Byzantinus 1502
Livy, 8.22.5	T	3, ix^r	
Libanius, *Argum. Dem.*, 1.1.1	GT and LT	3, ix^r	*Demosthenis Orationes* 1513

Book 1

Passage or work	Type of citation	Position in the *De Nola* (chapter and folium)	Possible printed or ms. source (only for the Greek works)
Livy, 23.45.10	T	4, ix^r	
Silius, 12.162–164	T	4, ix^r	
Livy, 7.22.5	T	6, ix^v	
Livy, 9.28.5–6	T	6, x^r	
Livy, 23.45.10	T	6, x^r	
Livy, 23.44.1–2	T	6, x^r	
Paulinus, *ep.* 32	M	6, x^v	
Paulus Diaconus, 15.17	R	6, x^v	
Livy 33.16	P	7, xi^v	
Cicero, *De off.* 1.10.33	R	10, xiiii^v	
Valerius Maximus, 7.3.4	R	10, xiiii^v	
Livy, per. 88	T	10, xv^r	
Cicero, *ad Att.* 13.8	T	10, xv^r	
Livy, 9.28	T	10, xv^r	
Galen, *De methodo medendi*, 5.12 (10.363–365 Kühn)	LTransl	11, xv^v, xvi^r	Galen 1500
Pliny the Younger, *ep.* 6.16.10–11	T	11, xvi^r	
Columella, *De re rust.* 10.133	T	11, xvi^r	
Solinus, 2.18	P	11, xvi^v	
Stephanus Byz., s.v. Βέσβικος, p. 165–166 Meineke	R	11, xvi^v	Stephanus Byzantinus 1502
Dio Cassius, 56.29.2	P	12, xvii^r	?
Dio Cassius, 56.31.2	P	12, xvii^r	?
Dio Cassius, 56.46.3	P	12, xvii^r	?
Suetonius, *Tib.* 40	P	12, xvii^r	
Suetonius, *Tit.* 8.4	P	12, xvii^r	
Pliny the Elder 21.53	P	12, xvii^v	
Livy 9.28	P	13, xvii^v	
Livy 23.44	T	13, xvii^v–xviii^r	
Livy 27.10	T	13, xviii^r	
Livy 23.15.8	T	13, xviii^r	
Livy 23.16	P	13, xviii^r	
Livy 9.28.5–6	P	13, xviii^r	

TABLE 1 *Classical sources quoted in the* De Nola *(cont.)*

Book 1

Passage or work	Type of citation	Position in the *De Nola* (chapter and folium)	Possible printed or ms. source (only for the Greek works)
Livy *per* 73	T	13, xviii^r	
Livy *per* 89	T	13, xviii^r	
Synesius, *ep.* 1.1–3	P	14, xx^r	*Epistolae* 1499
Aristotle, *Poet.* 1451b	R	14, xx^r	
Plutarch, *De audiendis poetis*	M	14, xx^r	Plutarchus 1509?
Stephanus Byz., s.v. Ἄτελλα ?	R	14, xx^r	Stephanus Byzantinus 1502
Philostratus the Elder, *Imag.*, Proem. 4–5.	R	14, xx^r	Lucian, Philostratus 1503
Cato, *De agri cult.* 151.1	R	14, xxi^r	
Cato, *De agri cult.* 135.2	R	14, xxi^r	
Strabo, 5.4.8 ?	R	14, xxi^r	Guarino's Latin translation?

Book 2

Stephanus Byz., s.v. Βεσκία, p. 166 Meineke	R	5, xxvi^v	Stephanus Byzantinus 1502
Paulinus, *ep.* 32, 10–17	T	12, xxxiii^v–xxxv^v	

Book 3

Apollonius Rhodius, 2.1131–1133	R	11, lvi^r	Apollonius 1496 ?
Homer, *Ilias*, 3.103–107; 264–313	R	11, lvi^r	Homer 1488 (Florence) or Homer 1504 (Venice)
Sophocles, *Aiax*	M	12, lvi^v	Sophocles 1502 (or Musuro's academic courses on Sophocles)

CHAPTER 2

Leone's Antiquarian Method and the Reconstruction of Ancient Nola

Bianca de Divitiis, Fulvio Lenzo

The importance of Nola in antiquity was well known to humanists thanks
to several literary sources which recorded it as the city where the Emperor
Augustus had died.[1] However, in the fifteenth century little of this glorious past
was visible, since the main ancient buildings had been destroyed or lay buried under several layers of earth. This chapter analyses how Leone, in the first
book of the *De Nola*, reconstructed the appearance of the principal ancient
buildings of his native city by combining the study of classical texts, the in-
spection of the material remains which were still visible and the examination
of ancient inscriptions. The origins of Nola went back to pre-Roman times,
but its fame as a city was associated with the Roman period, when it became
an important centre of the *Ager Campanus* thanks to its location half way between Rome and Apulia at the crossroads of the main consular roads, such as
the Appia, the Latina and the Popilia.[2] Nola began to be deprived of its ancient
materials already in late antiquity, when they were used for the construction of
the early Christian complex of churches in Cimitile, just outside the town, on
the site of the tomb of its first bishop, the martyr Saint Felix, greatly expanded
by his successor Paulinus.[3] Impoverishment increased with the incursions of
Alaric's Goths in 410 and then with the raids by the Vandals, when the city
was further expoliated of its antiquities.[4] Furthermore, as Leone points out
at the beginning of the first book, the peculiar character of the Nolan territory, squeezed between Mount Avella and Vesuvius [fig. 1], was the reason for

---

1 The main classical sources on the death of Augustus at Nola are: Velleius 2.123; Suetonius,
   Augustus, 98.10; 100.1–2; Tacitus *Annales*, 1.5.3; 1.9.1; Dio Cassius, 56.29.2; 56.31.2; 56.46.3; see
   see Bonghi Jovino, Niro 1993. Modern literature: Della Corte 1933–1934; Immarco 1994.
2 Leone 1514, bk. I, ch. 5, f. ix *verso*. For the ancient history of Nola, see Beloch 1890, 389–410;
   Quindici 1984; Savino 2005, 207–213.
3 Ebanista 2005; Ebanista 2007.
4 The devastation by Alaric is recorded by St Augustine of Hippo (*De civitate Dei* 1.10). The evolution of the city during late antiquity is unclear, but it is known that the vast territory was
   organised in *pagi* and *villae*. See Camodeca 2001; Savino 2005, 207–213.

many further episodes of destruction.[5] After the fall of the Roman Empire, the town was many times destroyed by the combined action of floods and volcanic eruptions and many times rebuilt over its own ruins, on a higher level and within a smaller area.[6] Historical events, natural disasters and subsequent reconstructions which occurred over the centuries, gradually buried the ancient urban fabric and erased the magnificent remains of the ancient city.[7]

Nola's leading role in the region was revived only from the mid-twelfth century when it became a fief of the Orsini family. The new rulers promoted a veritable urban renovation of the town by refashioning the cathedral, founding new churches, paving the main streets and creating public buildings, such as the new Seggio, where the members of the local elite could meet.[8] In the same years when Leone was still living in Nola, the city was under the rule of the count Orso Orsini (1461–1479), who attempted to transform the medieval aspect of the capital of his fiefdom into a Renaissance *all'antica* city, by promoting studies on the ancient Roman history of Nola, unearthing the antiquities which had long been buried, building a magnificent new palace and transforming the main square into a new *forum*.[9] Leone was an eyewitness of such an exciting period, and he later transferred his direct knowledge of the excavations and construction works into the *De Nola*, combining it with a new antiquarian awareness derived from his readings and his participation in the lively and stimulating humanistic ambience first in Naples and then in Venice.

---

5   Recent archaeological excavations have revealed that it was precisely on account of volcanic eruptions and floods, such as the one which occurred in 505, that the ancient city was reduced to occupying solely the site of its ancient forum and that the area to the southwest and the large public edifices were completely abandoned. See Sampaolo 1987.

6   Leone 1514, bk. I, ch. 1, ff. iiii *recto*–vi *verso*.

7   An inscription in the Seggio recorded the floods of 1424 and 1504: CVM VERNA DILVVIA IN CAMPIS TVIS NOLANAE VIDERIS / MORS MORA, VITA FVGA ERIT. / A.D. MDIIII AB RADICIBVS CICALAE / MONTIS VSQVE PALVDES A PORTA VICANTIA / AD QVINGENTOS PASSVS VERSVS VESVVIVM ILLINC TENSA / HINC LATA PISCOSA STAGNA NOLANOS / FERE OMNES FEBRE MALA AESTATE VNA NECARVNT / IDEM OCTOGESIMO ANTE ANNO VENISSE NOVIMVS. See Remondini 1747–1757, III, 204; Lenzo 2014, 179. For the floods affecting Nolan territory in modern times, see Fiengo 1988.

8   For the religious foundations promoted by the Orsini family, such as the Franciscan churches and monasteries of Santa Chiara (1359), San Francesco (1372), and later Sant'Angelo in Palco (between 1420 and 1436), see Leone 1514, bk. II, ch. 13, ff. xxxvi *recto*–xxxvii *verso*; bk. II, ch. 16, ff. xxxix *verso*–xl *recto*; Vincenti 1897; Buonaguro 1997; Di Cerbo 2008; Di Cerbo 2013, 1–28. For the Seggio of Nola, described by Leone 1514, bk. II, ch. 10, ff. xxx *recto*–*verso*, see Lenzo 2014, 178–179, 200.

9   For the Orsini palace, see Clarke 1996. For the urban renewal promoted by Orso Orsini, see de Divitiis 2013, de Divitiis 2016, and de Divitiis in this volume, 81–92.

The result is an exceptional antiquarian reconstruction of the physical appearance of the city during antiquity [fig. 2].

Reading the *De Nola* it is possible to follow in detail the method adopted by Leone to reconstruct the history and the phisycal appearance of ancient Nola. In both cases, Leone had recourse not only to literary sources, such as Livy, Pliny, Strabo and Silius Italicus, but relied abundantly on ancient inscriptions which he had examined personally and in most cases also transcribed. Inscriptions were used by Leone as documents integrated into the main text to support historical reconstructions or to identify ancient remains. An interesting example of how Leone used inscriptions as a source for the history of Nola appears in book 1, chapter 6 when, in order to describe the rise of the city and the alliance between the Nolans and Samnites in the third century BCE, he integrates and completes the account given by Livy (8.23, 25, 26) with the information he had gained from an only partly legible inscription which he found walled in the left pilaster of the southern city gate of Mirabella Eclano, a city of Samnite origin.[10] Later in the text, Leone again transcribes in full three inscriptions mentioning the names and status of many local leaders, which were still visible in Nola and in the surrounding territory, so as to demonstrate the brave and belligerent nature of the Nolans, who had even been capable of resisting Hannibal.[11] In the *De Nola*, inscriptions are regarded as an indispensable tool with which the history of the city could be reconnected with its ancient physical appearance and remains which were still visible could be identified.

On some occasions Leone uses vague clues as if they were conclusive evidence. This seems to be the case of the temple dedicated to Emperor Augustus, whose existence in ancient times was demonstrated by literary sources, such as Suetonius (*Tiberius* 40), Tacitus (*Annales*, 4.57.1) and Cassius Dio (56.43.3), but

---

10   Leone 1514, bk. I, ch. 6, f. ix *verso*. The inscription is CIL, IX, 175*. See Guarini 1814, 98–100.

11   Leone 1514, bk. I, ch. 14 f. xx *verso*. The inscriptions quoted by Leone are CIL, X, 1273; CIL, X, 1262 and CIL, X, 1258. The inscription CIL, X, 1273 in the sixteenth century was part of the façade of the cathedral of Nola, to the left of the main door, and is now preserved in the Seminario Diocesano di Nola; see Antonini, Mollo, Solpietro 1997, 28. The inscriptions CIL, X, 1262 and CIL, X, 1258 were visible in the palaeo-Christian church complex of Cimitile. Gianstefano Remondini (1747–1757, I, 96), misundertood the inscribed texts and believed that the location in Cimitile of the inscription CIL, X, 1262 could have led Leone to locate in Cimitile the ancient temple of Apollo. Such hypothesis is unfounded. On the contrary, Leone based the alleged location of the temple of Apollo on a passage from the life of Saint Felix; see Leone 1514, bk. I, ch. 14, f. xix *verso*.

of which there were no traces in the fifteenth century.[12] Leone records having read the inscription "Templum Augusti" on a almost erased slab found together with other "square marbles" ("marmora aliquot quadrata") in the area between the two amphitheatres and he uses this to locate the temple in the city [fig. 2].[13] It is not possible to ascertain if Leone forged the inscription (CIL, X, 174*) as Theodor Mommsen believed was the case, but he cannot have been unaware that the elements at his disposal were far from being a "sure proof" (*certo argumento*) in support of what he declares. After having established the position of the temple of Augustus, Leone uses this constatation to identify an ancient structure nearby, relying again on epigraphic evidence. Thus he uses the inscription relating to the *magister Mercurialis et Augustalis* Lucius Sattius [CIL, X, 1272; fig. 26] to argue that the temples of Augustus and Mercury were near to each other and therefore that the latter could be identified with other ancient remains not far from the chapel of the Salvatore.[14]

As Leone clearly states, together with the inscriptions, the antiquity and nobility of Nola and of its inhabitants were proved by the remarkable monuments which had once been built in the city. He then adds that the remains of ancient Nola belong to four types, listing the two amphitheatres, the tombs, the pavements of some temples and the foundations of many other buildings, some of which still lay under the ground. Among the various remains, the only antiquities which could be considered as certain were the funerary monuments and the two amphitheatres, which he had inspected during his youth [fig. 2, 11, 12, 20]. All the others remained circumstantial. Leone discusses the remains of ancient tombs (*tumuli*) at the west of the city, along the route to Naples, which probably correspond to the remains of two still surviving Roman funerary monuments, now called the "Torricelle".[15] In his text Leone does not specify the number of such structures, but on the map of *Nola Vetus* only two monuments are represented. The map shows them crowned, inconceivably, by onion-shaped domes, but in the text Leone gives us a more detailed and appropriate description. These Roman mausolea were round buildings covered

---

12   For the episode as reported by Leone, and his reference to classical sources, see Miletti 2016b.

13   Leone 1514, bk. I, ch. 8, f. xiii *recto*; bk. I, ch. 14, f. xix *recto*; see above, 20–21 note 31.

14   Leone 1514, bk. I, ch. 14, ff. xiii *recto* and xix *recto*. In the sixteenth century it was part of the façade of the cathedral of Nola, to the right of the main door, and it is now in the Antiquarium of the Seminario Diocesano di Nola; see Remondini 1747–1757, I, 96; III, 578; Antonini, Mollo, Solpietro 1997, 24, n. I. On the *Mercuriales* and their relationship with the *collegia* of Augustales in Roman times, see Farnoux 1987.

15   For the "Torricelle", see Remondini 1747–1757, I, 106; Beloch 1890, 405; La Rocca, Angelillo 1971, 80; Cesarano 2011, 55–58.

with a pointed conic roof, and their walls were constructed in brick (*structura lateria*) so tightly fitted together that they appeared to be solid ("tam apte compacta ut materia solida").[16] From Leone's descriptions it is also evident that the *mausolea* had already been stripped of their revetments to the point that only the inner walls could be seen, just as today. A more fanciful reconstruction of one of these monuments is visible in a pair of drawings, one a copy of the other, the original executed by Antoine Morillon [fig. 22] and the copy by the anonymous draughtsman of album *Destailleur B* in the Hermitage, both dated to the mid-16th century [fig. 21].[17] Leone also refers to many other Roman mausolea visible during his time in the countryside outside the town along the road from Avella to Stabia; these are also recorded by Renaissance draughtsmen and modern scholars, but the Nolan humanist does not give us any further information about them [fig. 24, 25].[18] Leone was not interested in recording all the ancient monuments in the surroundings of Nola; he selected the remains which served to describe and demonstrate the magnificence of his native city in the Roman era.

According to the humanist, among the antiquities which should be considered certain there were not only one, but two amphitheatres, one in brick (*amphitheatrum latericium*) and the other in stone (*amphitheatrum marmoreum*): this hypothesis, probably supported in Leone's mind by the examples of Pozzuoli and Rome, would have allowed Nola to surpass the nearby rival city of Capua, as well as many other cities across the Roman Empire. To make his argument more convincing, the author gives the dimensions of the two buildings, which according to him both had a diameter of fifty steps; the distance between

---

16   Leone 1514, bk. I, ch. 8, f. xii *verso*.

17   The drawing by Antoine Morillon, Eton College Library, *Topham Collection*, ms. B0.17.4, f. 12 *recto*, with the entry "Nole ex Albano lapide", has never been published before; for a description see Crawford 1998, 108. The drawing in the album *Destailleur B*, f.110[107]v, is published and described by Lanzarini, Martinis 2015, 159–160, plate XXXVIII, where however the monument is compared to the *Templum Mercurii* represented on Leone's map of *Nola Vetus*; for a more convincing identification with the "Torricelle", see Cesarano 2012, 55, 58.

18   Leone 1514, bk. I, ch. 6, f. x *verso*. A monument from Teglanum, an ancient centre near Palma Campania, south of Nola, is again represented both in the albums by the anonymous *Destailler B* (f. 115[112] *verso*) and by Morillon (f. 1 *recto*); see Crawford 1998, 97, 108; Lanzarini, Martinis 2015, 164–164 and plate XL (although with an erroneous identification of the location). For the identification of ancient Teglanum in the neighbourhood of Nola, see Beloch 1890, 408. Modern archaeological researches have discovered other Roman mausolea in the countryside round Nola near San Paolo Bel Sito and along the road leading to Camaldoli; see Beloch 1890, 405; Maiuri 1957; De Franciscis 1967, 197.

them he calculated as being approximately 360 steps.[19] On the basis of these measurements he was able to reconstruct the circumference of the ancient city, from 1,980 to 2,074 steps, according to what Leone writes.[20] In Leone's day only the one which he calls the "brick amphitheatre" was still partly standing.[21] An example of a contemporary representation of it can be found in a fifteenth-century parchment map of the territory surrounding Nola where it is shown, albeit in no detail, and clearly identified as "antico anphit" [fig. 11, 12].[22] However, Leone concentrates his attention above all on reconstructing the other amphitheatre, the "marble amphitheatre", of which he could only see parts of the foundations outside the fifteenth-century walls. Only the modern archaeological excavations undertaken in the 1990s have definitively established that the building described by Leone as a round "marble amphitheatre" was instead a semi-circular theatre.[23] Leone recounts how, during his infancy, the buried remains of the "marble amphitheatre" were originally recognizable only by a swelling of the earth ("Dirutum enim aedificium verrucam grandem quasi naturalem extuberavit"), and by two isolated arches which emerged out of the ground surface at some distance from each other.[24] Combining only this evidence with the measurements of the foundation blocks and with the supposed overall dimensions of the circumference of the plan, Leone concludes that the two arches were the remains of an amphitheatre made of marble, larger and considerably more magnificent than the other Nolan amphitheatre which he considered to have always consisted only of bricks. By comparing the unearthed remains of the marble amphitheatre with the two orders of the brick amphitheatre which were still visible, the author concludes that the original appearance of the "marble amphitheatre" consisted of three orders, as this type of amphitheatre "used to be built with no fewer than three orders of arches".[25] Leone explicitly compared the Nolan remains with those of the amphitheatre in Capua, which he regarded as less magnificent, having been executed with less expense and in less time. But he most probably had other examples too in mind. For instance, it is possible that Leone had become

---

19   Leone 1514, bk. I, ch. 8, f. xii *recto*.

20   Leone 1514, bk. I, ch. 8, f. xii *recto*.

21   For the amphitheatre of Nola, called by Leone the "brick amphitheatre", see Sampaolo 1991; Capaldi 2004–2005; Welch 2007, 240–245.

22   Archivio di Stato di Napoli, *Ufficio iconografico*, n. 65, Selva di Maddaloni; see Cesarano 2011.

23   Sampaolo 1991. That Leone's "marble amphitheatre" was more probably a theatre is a conjecture already found in Beloch 1890, 405.

24   Leone 1514, bk. I, ch. 8, f. xii *recto*.

25   Leone 1514, bk. I, ch. 8, f. xii *recto*.

aware of the existence of the three-order amphitheatres of Verona and Pola, in
Venetian territory, when he was living in Venice, in the very years he was writing *De Nola*. From the material evidence of the remains and from his knowledge of other examples of Roman amphitheatres, Leone redesigns in detail the
ancient monument, even to the point of being able to claim that the building
would have accommodated over thirty thousand spectators.[26] These fragments
of evidence also led Leone to represent a virtual reconstruction of the amphitheatre in the engraved view of *Nola vetus* included in the book [fig. 2, 11, 12].

It is particularly interesting to note how in his description of the ancient
building, Leone not only made a distinction between the white limestone
blocks of the foundation and the many polished fragments of marble of the
decorative elements, but was also able to identify the different origins of these
stones. Leone specifies that the limestone had been excavated from the local
quarry of Cicala or from a quarry in nearby Stabiae, while the other marbles
were "exotic" (*peregrina*) and came from Liguria and Cape Circeo in southern Latium.[27] One wonders if this meant that he saw fragments of black or
dark red marble, such as the *Portoro* or *Rosso Levanzo*—the two main marbles
of Liguria—and pieces of alabaster, the only stone that could be associated
with Mount Circeo. But even if Leone's identification of the provenance of the
marble stones he saw is mistaken, it is still worthy of note both that he possessed such kinds of uncommon technical knowledge and that he employed
it in archeological investigations. According to Leone, the limestone blocks of
the foundations of the "marble amphitheater" were reused by the count Orso
to rebuild his palace [fig. 28, 29]. In this context, his account of the "marble amphitheatre", rather than being a simple description or reconstruction of the ancient monument, should be seen as an authentic testimony to a historic event.
The humanist had been present during the exceptional discovery and excavation of the foundations of the ancient monument carried out in the 1460s by
the lord of Nola, count Orso Orsini, who redeployed the ancient blocks to build
the facade of his palace.

Just as in Rome in those very same years, so in Nola the study of an ancient monument took place simultaneously with its spoliation[28] Orso's excavations presented Leone with the opportunity to study and measure the ancient
remains, distinguishing the original parts from those added in the medieval
period, when the monument was used as a fortress. It is not by chance that

---

26   Leone 1514, bk. I, ch. 8, f. xii *recto*.
27   Leone 1514, bk. I, ch. 8, f. xii *recto*.
28   For Fra Giocondo's lamentation on the destruction of ancient monuments in Rome, see
      Koortbojian 2002.

the description of the amphitheatre in chapter 8 of book I is in effect completed in the account of the Orsini palace in chapter 9 of book II: a physical and
historical connection is thus created between ancient and modern Nola. With
the marble foundations of the amphitheatre and those in tufa stone of the
palace in front of his eyes, Leone gives us an exceptional account of the movement of antiquities from the archaeological to the building site; his testimony
to the stones in the original location and their new life in the Orsini palace is
unique. According to Leone, Orso found he was able to extract from the foundations so much marble that he changed the design of his palace and decided
to extend the use of the blocks, which he initially envisaged just for the corners
of the building, to the whole façade as well as other parts, such as the many pilasters and bases and all the cornices. Almost as if Orso wished to give a documentary proof of the important source of the stone he had used to build his
palace, an inscription from the "marble amphitheatre", i.e. the Roman theatre,
was placed in the lower courses of the façade.[29] By taking possession of these
remains, Orso privatized, so to speak, a monument which was central to the
city's identity, but on the other hand, by bringing the ancient blocks to light,
he also restored to local citizens the material and visible presence of what was
Nola's most important ancient building. There is an extra dimension to Leone's
account of the move of the ancient stones to the new building site; the ancient
glory of Nola is being transferred to new representative buildings such as the
Orsini baronial palace, which he significantly calls a *Regia* [fig. 28], echoing the
distinction made by Leon Battista Alberti between the magnificent residence
of the prince, and the castle of medieval origins devoted to military functions.

From Leone's account it is clear that excavations were going on not only
on the site of the amphitheatre but also in other locations [fig. 2, 11, 12]. It was
during this "campaign" that other ancient remains emerged from the subsoil
two hundred steps in a south-easterly direction from the presumed temple of
Augustus, which Leone identified as the temple of Mercury [fig. 14].[30] As in
the case of the amphitheatre, this excavation too gave rise to the opportunity
to study the monument while its limestone blocks were moved to the building
site of the palace of Carlo Carafa in Naples, where they are still visible today
[fig. 31].[31] Leone writes that count Orso was, in his magnanimity, in the habit

---

29   Parma 2015; de Divitiis 2016.

30   Leone 1514, bk. I, ch. 8, f. xiii *recto*. Leone states that since the foundations were in stone,
      they could belong only to a temple and not to a private house, which would have had a
      simple brick foundation.

31   The palace, converted into the Collegio of the Gesù Vecchio at the end of the sixteenth
      century, belongs today to the University of Naples Federico II. The provenance of the

of giving ancient stones that were left over to many private citizens.[32] It is possible that among them there was also the long Doric frieze used to decorate
the base of the façade of the Albertini palace [fig. 30], a building dating to the
same years as the Orsini palace and standing just opposite it. Modern scholars
suggest that the metopes and triglyphs originally belonged to a public monument of ancient Nola, most probably a portico resting on free columns.[33] It
is tantalizing, therefore, to speculate that such remains might have appeared
during Orso's renovation works in the main square in front of the cathedral,
which rested on the site of the forum of Roman Nola.[34] However, although
this long Doric frieze sculpted with military trophies was (and remains today)
the most striking example of antiquity reused in modern Nola, Leone does not
write about it.

Instead, relying on material evidence, on the excavations which were in
progress and on his literary knowledge, Leone tried methodically to shed light
on the buried material antiquities of Nola. The most vivid image of Leone's
tireless efforts to rediscover the antiquities of Nola and of the discoveries he

---

stone blocks of its base was misinterpreted by Pietro Summonte in 1524, when he wrote
that they came from the ancient Nolan palace in which the emperor Augustus had died;
see Nicolini 1925, 174, 177, 281–282; Pane 1975–1977, I, 63–71. Giovanni Antonio Summonte
1601, 298, instead, wrote that the blocks came from the temple of Augustus in Nola and,
on the palace in Naples, adds that "ridotta la fabrica del Palaggio intorno à palmi 10 sopra
terra, mancando di vita [Carlo Carafa] rimase il Palaggio imperfetto fin'all'anno 1557 nel
cui tempo, venuti in Napoli i Preti Gesuini vi edificarono su quel principio la lor Chiesa,
al presente chiamata il Collegio de' Gesuini". See also Remondini 1747–1757, I, 95–96;
Beloch 1890, 404. It is interesting to note how for non-Nolan humanists, all the archaeological remains of the city were to be related, in some way, to the only well-known episode
from its ancient Roman history, the death of Augustus.

32   Leone 1514, bk. I, ch. 8, f. xii *recto*.

33   Polito 1998, 140–141, believes that the frieze belonged to a funerary monument, but
      Capaldi 2005, 70–80, argues that the metopes and triglyphs must have been part of a free
      entablature, and therefore concludes that it belonged to a portico on columns.

34   The origins of this frieze have not yet been determined. Since fragments from the same
      ancient monument were reused in the twelfth-century bell tower of the cathedral and
      in the fifteenth-century palace of the Albertini, it is probable that this ancient building
      was subject to excavation in both periods. Therefore it is probable that it was located in
      the vicinity of the cathedral, and came to light when the cathedral was built, as well as
      during the renovation of the square in front of it which was undertaken by count Orso
      Orsini at the end of the 15th Century. See De Divitiis 2016, 38–42. Leone 1514, bk. II, ch. 11,
      f. xxxiii *recto*, refers to the fragments of the frieze in the cathedral bell tower as "marbles
      which attest to the poetry and military art of their age" ("marmora quaedam ordine locata
      sculptaque sunt tum poesim tum militarem artem aetatis suae testantia").

saw being made in front of his eyes comes in his repeated descriptions of how
he felt antiquity under his feet and observed the ruins of ancient constructions,
such as houses, temples and theatres making the ground surface swell ("solum
extumescit ob antiquarum aedium ruinam") [fig. 2].[35] It was through the investigation of one of these many swellings in the soil that he recognized not
only the amphitheatre but also the gate of the ancient city walls.[36] We also find
Leone often looking down on the sunken pavements of the buildings which
had been buried by the rising ground level ("humi depressus comperitur").[37]
This is the case of the churches of Santi Apostoli and of San Felice in Piazza,
where the lowering of the floor indicated they had been founded in remote
times.[38] The paving of the latter church was recognized by Leone as the remains of a temple, even though the absence of futher evidence meant that he
gave up the attempt to make a more specific identification.[39] This is also the
case of a great ancient wall, the location of which Leone is keen to get right [fig.
9]. He writes that it was situated forty steps north of the temple of Jupiter, and
attached to the houses of the Cesarino, Mazzeo, Frezza and Tango families.[40]
However, Leone is unable to understand the origins of this solid wall of great
width, and he wonders if it belonged to a large ancient house or to a temple
or even to the late antique or early medieval city wall.[41] Providing what is a
unique image of the excavations in progress as they being carried out at the
time in Nola, Leone records in the text and on the map of *Nola vetus* several
unidentified remains which had been excavated in various areas of the city,
such as walls or other types of traces which were found during the excavations
near the gate which led to the ancient *Via Abellae*, as well as those sites where
excavations were underway.[42]

---

35   Leone 1514, bk. I, ch. 8, f. xiii *recto*.
36   Leone 1514, bk. I, ch. 8, f. xiii *recto*: "Quinetiam ab Augusti templo recta in austro pro-
     ficiscenti trecentos passus fit obvius locus quidam extumescens ob antiquae fabricae
     fragmenta, quas reliquias murorum antiquae urbis fuisse mensus atque rotunditas urbis
     ipsius ostendunt, quinetiam illic portam nunc cum turribus patuisse [...]. Ab amphithe-
     atro praeterea marmoreo recta eunti versus Vulturnum ventum ad centum passus locus
     extumescens alter occurrit, ubi murus urbis stitisse[sic] eadem ratione creditur atque si-
     militer porta cum turribus quae Sarnum Nuceriamque euntibus iter praebebat".
37   Leone 1514, bk. I, ch. 8, f. xii *verso*.
38   Leone 1514, bk. II, ch. 11, f. xxxiii *recto*; bk. I, ch. 8, f. xii *verso*; bk. II, ch. 13, f. xxxvi *verso*.
39   Leone 1514, bk. I, ch. 8, f. xii *verso*–xiii *recto*.
40   Leone 1514, bk. I, ch. 8, f. xiii *verso*.
41   Leone 1514, bk. I, ch. 8, f. xiii *recto*.
42   Leone 1514, bk. I, ch. 8, f. xiii *recto*; bk. II, ch. 1, f. xxii *recto*.

Leone understood that in Nola, more than in any other centre of Southern Italy, antiquity could be investigated by studying the subsoil of the modern city. This can be inferred from his descriptions of different kinds of ancient structures which he inspected underground. He mentions for example the presence of an ancient underground porch ("subterranea porticus, antiqum [sic] opus") between the two religious foundations of Sant'Angelo in Palco and San Giovanni Cesco near Nola, specifying that some architects used to call such structures *cryptoporticus* ("quam cryptoporticus architecti quidam vocant"); he adds that they were used for keeping warm in the winter and cool in summer and that they could be built either for public or for private use.[43] It is probable that in his reference to architects, Leone is alluding to Leon Battista Alberti, who deals with the *cryptoportici* when explaining the different kinds of vaults,[44] since Vitruvius did not use the word, which instead appears in Pliny's description of his Laurentine villa, while in Suetonius' life of Caligula the word used is *crypta*.[45] The word was more familiar to modern authors after Nicolò Perotti († 1480) explained, in his *Cornu copiae*, that the *cryptoporticus* was an underground arcade used in summer, a description summarised in Francesco Maria Grapaldi's *De partibus aedium* of 1494, and almost exactely replicated by Ambrogio Calepino in his *Dictionarium* (1502).[46] Even before the term *cryptoporticus* became familiar, the use of such a structure was well-known in Angevin Naples, where the *Cronaca di Partenope* describes them as "underground summer rooms" ("habitaciune subterranee de la extate").[47]

Leone was fascinated also by the underground aqueduct (*subterraneum aqueductum*) which carried water from the Serino spring to the city of Naples, in which the water flowed through deep tunnels ("per altos ad id factos cuniculos").[48] However, in spite of his interest in Roman underground structures, he surprisingly pays minimal attention to the existence of ancient

---

43 Leone 1514, bk. II, ch. 6, f. xxviii *recto*.
44 Alberti (ed. Orlandi) 1966, bk. III, ch. 14, 240–241.
45 On the interpretation of the term *cryptoporticus* see Zarmakoupi 2011.
46 Perotti (ed. Charlet et alii) 1989–2001, bk I, epigram 2, vol. 2, paragraph 155, 65: "Sunt autem cryptoporticus subterraneae porticus cum amplis fenestris quibus refrigerandi g.ra per aestatem utimur". Perotti's monumental *Cornu copiae* was printed first in Venice in 1489, and then reprinted several times, enjoying great renown. Grapaldi 1494, bk. I, ch. 1, unnumbered pages, *sub voce Porticus*: "Subterraneas autem Crypropoticus appellant". Calepino 1502, *sub vocem*, unnumbered pages: "Crypto porticus nomen compositum ex graeco latinoque et significat subterraneas porticus cum amplis fenestris quibus refrigerandi gratia per aestatem utimur".
47 *Cronaca di Partenope* (ed. Kelly) 2011, 179.
48 Leone 1514, bk. I, ch. 1, f. v *recto*.

aqueducts within the territory of Nola. This is the case of the ancient aqueduct which served Nola, the remains of which still existed in the nearby city of Lauro (*Laurinum*), which Leone describes in the first book as part of his reconstruction of the *Ager Nolanus*, just after having discussed the hydrography of the area. Leone describes the aqueduct as being very small (*parvus aquaeductus*) and adds that it was one of the two watercourses which flowed in the territory around Nola, together with a modest stream coming from Mount Avella.[49] The description of the aqueduct is part of Leone's attempt to refute the anecdote reported by the annotator Aulus Gellius, according to which Virgil erased Nola's name from a verse of his *Georgics* (2.225) when the Nolans did not allow him to run public water into a farm belonging to him. At the time of Leone, the anecdote was at the centre of an intense Renaissance debate involving major humanists of the day, like Pontano and Sannazaro, who respectively took up positions for and against Nola. Leone's work would also have been the means by which news of the debate reached Erasmus.[50] Leone rejects the story as untruthful, arguing that the modest dimensions of the aqueduct and the weakness of the water flow, which could be used only for ornamental purposes, were in themselves the demonstration that it was impossible to run water from the aqueduct without letting it become totally dry. Leone also asserts that the little stream with its source on Mount Avella could not have been the stream in the anecdote as it had always been public, and there would have been no point in Virgil requesting to use its water since whoever needed to do so was free to channel water from it into his own fields.[51] Thus Leone's description of the ancient aqueduct should be interpreted in the light of his attempt to defend Nola's reputation. Considering his expert interest in ancient monuments, Leone's insistence on the modest dimensions of the watercourse from Laurino and its purely ornamental function implies a biassed interpretation, motivated by a wish to eliminate any indication that could allow a possible identification with the watercourse the water from which had supposedly been denied to

---

49   Leone 1514, bk. I, ch. 1, f. v *verso*: "Ager autem ipse caret flumine omne genus atque rivulo et fonte. Solus unus rivulus (ut dictum est) secundum Nolanum campum a sinistro cornu Abellae defluit hyberno verno tempore, aestate vero arescit adeo ut vix in suprema valle atque etiam tenuissimus inveniatur. Quinetiam extant vestigia quaedam parvi aquaeductus in ea via quae Laurinum fert". For the Roman aqueduct of Nola see Beloch 1890, 408; Ohlig 2001.

50   Miletti 2016a. See this article also for Leone's ignorance of the description of the Nolan aqueduct by Paulinus (*carm.* 21). For this latter episode, see Luciano et alii 1980; Savino 2005, 242.

51   Leone 1514, bk. I, ch. 2, f. viii *recto*: "Nulla itaque aqua negari potuit a Nolanis, nam Abellana non erat negata, Lauriniana non erat petita".

Virgil. Leone's account is even more striking if we consider how, just a few decades later in 1560, the remains of the same ancient aqueduct were described in detail in the survey carried out by Neapolitan engineer Pietro Antonio Lettieri as an integral part of the structure supplying water for Naples.[52] Leone's problematic relation with the aqueduct is confirmed by the fact that he only just indicated the remains of its branch in Naples in the map of the *Ager Nolanus* [fig. 1, 5, 6], identifying it with the river Sebetus; the aqueduct does not appear at all in the *Nola Vetus* [fig. 2], where—as we have seen—he was eager to indicate every surviving vestige of antiquity.[53] Leone's bias probably also explains the omission from the text of a well-known ancient fragmentary inscription (CIL, X, 1285) [fig. 27], placed in a prominent position in the episcopal precint, which stated how a certain amount of water from the ancient aqueduct had been granted to a private individual.[54] The inscription, which had also been copied by the antiquarian and humanist Fra Giocondo at the end of the fifteenth century, would have explicitly contradicted Leone's strategically built arguments against Gellius's notorious accusation of the Nolans' lack of generosity. In the context of *De Nola* the description of the aqueduct is a telling case where a literary source becomes the reason to diminish the importance of ancient remains, rather than emphasise it.

To sum up, Leone's attempt to reconstruct and describe the physical appearance of ancient Nola responds to the main aim which underlies the whole book, to bring to light the ancient magnificence of his native city and at the same time to demonstrate how its noble character in antiquity lay at the origin of the nobility of the contemporary city and its inhabitants. In order to pursue this refined strategy Leone combined his humanistic knowledge of ancient authors, with uncommon antiquarian and architectural skills; these allowed him to merge the accounts of the ancient texts he cited with his analysis of the material evidence of the monumental remains and inscriptions. Several interesting aspects emerge from his approach to antiquities, which make his descriptions exceptional for the period. Leone attributes considerable importance to material archaeological remains: in so doing, Leone attempts to reconstruct the original appearance of the monuments which he saw, comparing

---

52   Lettieri [1560] 1803, 405. On the text by Lettieri, see Fiengo 1990, 19–38; Cosimi 2008.

53   See Lenzo in this volume, 63–71.

54   At the end of the fifteenth century, Fra Giocondo copied the inscription on the exterior wall of the church of Santi Apostoli ("in exedra Sancti Apostoli versum viam publicam"), while in the eighteenth century Remondini 1747–1757, I, 33, 117–118 described it near the Seggio; now it is preserved in the Antiquarium of the seminary; see Antonini, Mollo, Solpietro 1997, 24–25, n. III; Lenzo 2014, 121–122, 179.

them with other similar but better preserved buildings. He also tries to ground his reconstructions in an impressive quantity of numerical data, such as calculations of the possible dimensions of the original buildings, of the number of spectators the amphitheatres could hold and the distance measured in steps between different sites. He also dedicates much attention to the materials used in buildings, noting if remains were in stone, marble or brick and noting the different possible geographical provenances of coloured marbles, as we have seen in the case of the marble amphitheatre. In some cases, as in the temple of Mercury, Leone felt able to identify through the fragments of an inscription not only the function but also the dedication of a—in this case sacred—building. When instead he could not connect the material remains with epigraphic evidence or a literary source, as in the case of the great ancient wall north of the temple of Jupiter, he limited himself to recording the data he had collected. He was able to read the different layers in the same site, distinguishing what he thought to be the original parts from ones which had been added in subsequent periods. By providing an accurate account of what he could actually see with his eyes, Leone creates a vivid picture of the excavations which had been underway since the time of his youth and which, thanks to the patronage of the count Orso Orsini, had gradually unearthed the antiquities of the city which had long remained buried. These excavations, which were accurately recorded on the map of *Nola Vetus*, had given Leone an exceptional opportunity to study the monuments and develop their historical reconstruction in material evidence which came to light. Furthermore, Leone's book provides a unique account of the transfer of antiquities from the archaeological sites to sites where modern buildings were being constructed in Nola and in Naples. Indeed, it is this image of the city as a entire building site, where excavations went on in parallel to the construction of new edifices, such as the Orsini palace, which may explain Leone's constant interest in the subsoil and in the remains of ancient buildings which lay below the surface of the city with all its contemporary activity. It is interesting that he devotes more attention to what was not visible, such as the marble amphitheatre or the aqueduct, than to what was still standing, examining the remains more for what they could tell, rather than simply describing them as they were. Drawing on his antiquarian and archaeological researches, Leone constructs a picture of Nola which shows the ancient and contemporary magnificence and nobility of the city and of his inhabitants.

CHAPTER 3

The Four Engravings. Between Word and Image

Fulvio Lenzo

In the introduction to the *De Nola*, Ambrogio Leone declares that his book is composed of history and of images. Images, he adds, are endowed with an almost divine force since they are able to convey the real appearance of far-away things as though they were near and in front of one's eyes.[1] And indeed one of the most innovative features of the *De Nola* is the presence of four engravings, which, as Leone states in the introduction, were the result of his collaboration with the painter Girolamo Mocetto.[2] These plates represent, respectively, the southern territory of the Terra di Lavoro (*Ager Nolanus*) [fig. 1], the old city of Nola in Roman times (*Nola Vetus*) [fig. 2], the dimensional comparison between the old and the new town (*Figura praesentis urbis Nolae*) [fig. 3], and finally the modern town as it appeared at the beginning of the 16th century (*Nola Praesens*) [fig. 4].[3] This chapter will discuss the engravings of the *De Nola*, analysing the particularities of each plate, trying to identify their visual sources and focusing on their specific meaning in relation to what is written in the text, since it is only by comparing the images with the words employed by Leone that it is possible to understand their meaning. I will also discuss the attribution of the plates in trying to demonstrate that the main role in conceiving and most probably also in drawing them must be ascribed to Leone himself.

The order of the plates as they are inserted among the pages of the book strictly follows the structure of the text, which proceeds from the general to the particular, and from the past to the present. The illustrations are very different

---

1 Leone 1514, *Praefatio*, f. ii *verso* (see below, Appendix 2, § 6): "Imaginis enim vis atque actio divina nimirum est, quippe quae veram suamque rei cuiusque faciem agit et offert atque absentem tam exacte praesentem ostendit ac si foret in conspectu".

2 Leone 1514, *Praefatio*, f. ii *verso* (see below, Appendix 2, § 5): "... ipse imagine atque aspectus proprii similitudine tanta produxi in medium adiutus opera Hieronymi Moceti pictoris, ut oculis omnium atque ubique terrarum perquam facile possit esse conspicua". On this passage see also Miletti, in this volume, 16–17.

3 On the maps in the *De Nola*, see Almagià 1913, II, 319; Hind 1938–1948, V, 170–171; VII, tavv. 734–735; Manzi 1973; Weiss 1958, 178–179; Avella 1990; Carillo 1996; Ruggiero 1997. See also Cantabene 2006 and Miltenov 2009, who, however, both wrongly discuss the eighteenth-century copies of the engravings published by Pieter van der Aa (1729, plates 60–63) as if they were the original ones printed in 1514.

one from another not only in their visual aspects but also in the method of representation they adopt and their position in the volume. Three plates (*Ager*, *Nola Vetus* and *Nola Praesens*) are maps, oriented with the north at the top, while the *Figura praesentis urbis* is made up of a number of geometric figures. The *Ager* [fig. 1] is printed on a large single sheet, folded in half between two blank unnumbered sides immediately after leaf iii *verso* and before iv *recto*. The other plates are instead printed on sheets of the same format as the text of the book and, in the cases of *Nola Vetus* [fig. 2] and *Figura Praesentis urbis* [fig. 3], they are also included in the sequence of numbered leaves, respectively as leaves x *recto* and xxiii *verso* It should furthermore be noted that in several copies of the *De Nola* the first and/or the fourth plate are printed in red ink or in a shade of green.[4] This use of different colours perhaps reflects a wish to distinguish or personalise individual copies for a particular category of acquirers, in the same way, for example, that Aldo Manuzio occasionally printed a few copies of his editions on vellum or on blue paper.[5]

The text of the *De Nola* was written in order to be self-sufficient and it contains many detailed descriptions which enable readers to draw in their own minds exact images of landscapes and buildings without the need of illustration. Leone's main model for a book where images are conveyed by words was certainly the *De re aedificatoria* by Leon Battista Alberti, even though he may also have been influenced by other of Alberti's works, such as the *Ludi mathematici* and the *Descriptio Urbis Romae*. Alberti had planned his books in response to conditions which prevailed before the invention of printing, when texts could be transmitted but accompanying drawings could easily be lost, as had occurred with the first-century manuscript of Vitruvius which had come down to the fifteenth century lacking its images. In the same way Leone renounced the use of images in his book, preferring to describe architectural details by comparing them to geometrical forms or alphabetical characters, and to translate topographical maps from drawings into a series of coordinates and numbers.[6] Following Alberti, Leone describes in detail every feature and aspect of the town, such as its present-day form, its territory and its visual appearance in antiquity. He also furnishes the dimensions of ancient and modern buildings, measuring distances and describing shapes, suggesting parallels

---

4   For the presence of coloured engravings, see Hind 1938–1948, part II, V, 170–71, nos. 19–22; Mortimer 1974, II, 371–373, no. 255. I wish to thank Seth Fagen and William Stenhouse for having called my attention to this peculiarity of the book.

5   Nuvoloni, Parkin, Sachet 2016, 86–89; Toniolo 2016.

6   On the mutual influences between text and pictures, with special reference to Alberti's *Della Pittura*, see Spencer 1957. For Alberti's "iconophobia", see Carpo 1998.

with geometrical figures or with even more familiar objects such as a monk's habit to explain what the groundplan of the cathedral looks like or a man's half footprint to describe the way the city is laid out.[7] However, Leone also departed from his Albertian model by adding four engraved plates to his book. The fact that the text does not in itself need any illustration makes the insertion of the plates even more significant.

By 1514 the production of illustrated books was an already established tradition in European printing and publishing, especially in Venice. Here Leone must have admired the *Hypnerotomachia Poliphili*, printed by Aldo Manuzio in 1499.[8] Another two extremely interesting examples of the use of illustration were the new edition of Euclid's *Geometria* by Luca Pacioli (1509), and Fra Giocondo's edition of the *De architectura* by Vitruvius (1511), both printed in Venice during Leone's residence in the city.[9] As Pacioli himself tells us, Leone attended, together with Fra Giocondo, his *lectio* on the fifth book of Euclid's *Elements*, held on the 11th August 1508 in the church of San Bartolomeo at Rialto (fig. 63).[10] We can therefore assume that Leone knew the books by Pacioli and Giocondo and was probably in personal contact with their authors and would have had the opportunity to discuss with them the question of printing an illustrated book. However, all these books are illustrated by woodcuts and not by engravings, as the *De Nola* is. It was perhaps the wish to change the colour of the plates—which, as we have seen, were printed in black, red and greenish ink—that induced Leone, or possibly the printer Giovanni Rosso, to prefer engravings from metal plates—probably of copper—rather than woodcuts as was usual at that time.[11] However it should be noted that engravings presented other advantages. If woodcuts had to be printed using the same flat-bed press on which texts were printed, engravings, instead, could be carried out using a

---

7   Leone 1514, bk. II, ch. 2, f. xxii *verso*: "Proinde dimidio humani calcis vestigie, quod a talo est, urbs haec spectatur similis". Leone 1514, bk. II, ch. 11, f. xxxi *verso*: "Quamobrem hae duae aedes figuram T literae praesentant, cui si cellam superaddideris, tunicam cum cucullo ac extentis manicis finxeris".

8   For the *Hypnerotomachia Poliphili*, see the comment by Lucia Ciapponi and Giovanni Pozzi in Colonna (1499) 1980, II;, and Szépe 2016. For the derivation of Mocetto's engraving representing the *Metamorphosis of Amymone* from a plate of the *Hypnerotomachia Poliphili*, see Romano 1985, 47–51.

9   For Pacioli see Benzoni 2014. For Giocondo's edition of the *De architectura* by Vitruvius, see Pagliara 1984; Pagliara 2001; Salatin 2012; Pagliara 2014; Salatin 2014.

10   The list of participants is given by Luca Pacioli in Euclid (ed. Pacioli) 1509, f. 31 *recto*. See Danzi 2005, 25–25, note 38. See below in this chapter, note 50.

11   Generally engravings were printed from copper plates, however sometimes brass, tin and pewter plates were also used; see Landau, Parshall 1994, 23–24.

much smaller and cheaper roller press.[12] Therefore, woodcuts have to be printed in the printers' workshop at the same time as the book (indeed, as part of
the printing of the book), whereas engravings could be executed independently from the text, in other places, generally by the artist in his own atelier, and
at a different time, and then inserted into the foliation of the volume during
binding.

When the *De Nola* was printed, the number of images and their position in
the volume must already have been decided, as is demonstrated by the pages
left blank for the insertion of the engravings. On the other hand, one can argue
that when Leone began to write the *De Nola* he intended to publish a unil-
lustrated book. It was for this reason that he gave so much importance to detailed descriptions filled with visual comparisons, thus allowing the reader to
reconstruct virtual images in his mind. Apart from a few details in the *Ager
Nolanus*, the plates do not give any additional information to what one can
read in the text. But, as we will see, they notably reinforce its message, and at
the same time they transform the *De Nola* into a more elegant and modern
publication. This strange relation between text and images in the *De Nola* also
reflects the influence of the "Ptolemaic fever" or *furor geographicus* of the fifteenth century.[13] Ptolemy had originally written his *Geographia* in the second
century CE without illustrations, but providing careful descriptions and above
all exact series of coordinates and numbers, a method which allowed him to
describe most of the lands which were known about when he was alive. Even
if the text of the *Geographia* does not need any illustration, Renaissance editions were generally illustrated with maps and city views. In addition, Ptolemy
himself had distinguished between large-scale mapping—the *geographia*—
and the smaller and more detailed representations of places, the *chorographia*; the execution of the latter was reserved for draughtsmen or even painters
(Ptol. *Geogr.* 1.1.1).[14] Leone was surely aware of Ptolemy's distinction when he
specified that in illustrating the *De Nola* he had been assisted by "the painter"
Girolamo Mocetto.

---

12   Landau, Parshall 1994, 3, 12–14, 23–30; Landau 2016.

13   Nuti 1999; Dalché 2007; Kent, Elam 2015, 71.

14   The original Greek text does not mention painters, but in the 1409–10 Latin translation by
        Jacopo Angeli da Scarperia one can read "unde corographia pictura eget; nullusque eam
        recte componit nisi homo pictor". See Nuti 1999, 90; Kent, Elam 2015, 73.

1 The Territory: The *Ager Nolanus*

The only engraving signed by Mocetto as "HIE. MOCE" is the first, representing the *Ager Nolanus* [fig. 1, 5, 6]. Two states of this engraving are known, the later showing a little plant near the signature [fig. 5, 6]. The purpose of the map is to show the close connection of the town with the surrounding countryside. Here we find all the mountains (*Vesuvius, Abella, Pausilipus, Sarnus, Gaurus*), the hills (*Gecala, Vescianus*), the rivers (*Clanius, Sebetus, rivulus Abellanus*, and an unnamed *rivus* near Naples), the marshes (*Paludes, Palus Neapolitana*), the sulphur springs called *Mephitis*, and the artificial channel *Lagynos*, all described in book I, chapter 1.[15] Nola occupies the centre of the area, between Vesuvius and Mount Avella. In the text the two mountains are described as contrasting natural elements: Vesuvius was dry and associated with heat and fire, whilst the Avella mountain is described as often covered by snow and traversed by torrential waters.[16] Although not described in the text, the presence of an extensive stone quarry on the slope of mount Avella is instead heavily emphasized on the map by being depicted as a large black patch. The map contains representations of many towns and natural elements, but in contrast to 'geographical' territorial maps derived from Ptolemy, in the *Ager Nolanus* we can find in a single representation not only geography and chorography mixed together, but also history. The map of the *Ager* is the graphic translation of the antiquarian method developed by Biondo Flavio, Pomponio Leto, and other humanists, and here adopted also by Leone: first he surveys the topography of the territory, and then he adds historical information derived from classical literary texts; he even represents the no longer surviving cities of Pompei and Ercolano and specifies the areas inhabited in antiquity by the *Samnites* and the *Hirpini populi*. In this respect, the map of the *Ager* can be regarded as a cultural product of the Neapolitan humanistic milieu of the late fifteenth century, and reflects the studies on southern Italy in antiquity, even in pre-Roman times, carried out by Giovanni Pontano, Jacopo Sannazaro, and Antonio de Ferraris Galateo. It is also interesting to note that just a few years before Leone's books another 'Neapolitan' living in Venice, Bernardino Silvano da Eboli, in the maps of European countries included in his new printed edition of Ptolemy's *Geographia* (1511) indicated in red ink the names of the peoples

---

15   For the spring of Mefite, also called Mofete or Mofito, and its relation with the marshes of northern Campania, see Fiengo 1988, 2, 5, 10, 11, 17, 18, 31, 32, 50, 55, 82, 90.

16   Leone 1514, bk. I, ch. 1, ff. iiii *recto*–vi *recto*.

who had once lived in the different ancient regions.[17] A similar commixture of geography, archaeology and history had already appeared in a rare set of four parchment maps made in Naples in the late years of Aragonese rule: they showed, respectively, the territory between Maddaloni and Nola [fig. 17, 18], Lower Latium, the islands of Ischia and Procida, and Cape Gargano in Apulia.[18] Here, together with the orography of the sites and the names of the urban centres, we can also find ancient Roman buildings depicted in what was presumed to be their original appearance, such as the amphitheatre of Nola [fig. 19], as well as information deduced from literary sources, as in the case of the river Clanius, identified with transcriptions of the verses by Virgil which referred to it (*Georg.* 2.225) [fig. 17].[19]

In the same way, the engraving of the *Ager Nolanus* [fig. 1] indicates the ancient history of the region. The former site of ancient Palepoli is shown north of Naples [fig. 5, 6], and, further away, one can see the *campus Romanus*, in other words the site conquered by the Roman senator Quintus Fabius taking advantage of the quarrel between Naples and Nola, as Leone recounts in the text quoting Cicero (*De Officis*, 1.10,33) and Valerius Maximus (7.3,4).[20] The most interesting detail of the *Ager Nolanus* is however the representation of the city of Naples [fig. 5, 6], by means of which Leone pays a tribute to the humanistic image of the city derived from the discussions which developed in the last decades of the fifteenth century among the scholars of the *Accademia Pontaniana*. Leone himself had been closely connected with the pre-eminent members of the academy, namely Giovanni Pontano and later Jacopo

---

17   Ptolemy (ed. Silvano) 1511; the book, dedicated to Andrea Matteo Acquaviva duke of Atri, is the first illustrated edition of Ptolemy's *Geographia* showing a map of America. A wonderful manuscript copy of Ptolemy's *Geographia* by Silvano is in ms. Par. Lat. 10764. See Blessich 1897, 41–47; Conti 2009.

18   Archivio di Stato di Napoli (henceforth ASNa), *Ufficio Iconografico*, nn. 64–67. The four parchment maps are the only surviving parts from a more general survey of the entire Kingdom; the overall responsibility for the project has been attributed to Giovanni Pontano. Copies on paper of other now undocumented parchment maps were made in 1767 and are now partly in the Archivio di Stato di Napoli (*Piante e disegni*, cart. XXXI, nn. 15, 19, 20, 22, 23), and partly at the Bibliothèque Nationale de Paris (*Cartes et Plans*, Ge.AA.1305/1–7). However it is impossible to assess how reliable these 18th-century drawings are as guides for the reconstruction of the original fifteenth-century maps. See Valerio 1993, 36–44; Valerio 1993b; Iuliano 2002; Valerio 2007, 945–951; Jacazzi 2008; La Greca, Valerio 2008; Cesarano 2012.

19   ASNa, *Ufficio Iconografico*, 65. The passage by Virgil on the river Clanius had been quoted also by Biondo (ed. White) 2005–2016, II, 302.

20   Leone 1514, bk. I, ch. 10, f. xiiii *verso*.

Sannazaro.[21] It is not surprising that he makes no reference to them in the text when writing of Naples, since he rarely acknowledges his debts towards modern authors, preferring to stress the references to classical sources.[22] The urban fabric of the city of Naples is symbolised by the cathedral and the walls, part of which had been rebuilt by Alfonso II, a deed Leone himself praises in the *Praefatio* of the *De Nola*.[23] Alongside the wall one can recognize Naples' five castles: Castel Nuovo to the west, Castel Capuano to the east of the city, Castel Sant'Elmo on top of the *Pausilipus Mons*, and Castel dell'Ovo jutting out towards the sea; also the tower of San Vincenzo and the Angevin jetty are depicted jutting out from the coastline. The road which emerges from the ancient tunnel known as the *Crypta Neapolitana* can be seen next to the signature of Mocetto, while on the other side of the hill there is a church, possibly that of Santa Maria di Piedigrotta.[24] A row of trees leads from this church to Castel Nuovo: it is the Chiatamone or Platamone, the area alongside the sea between the centre of the city and Mergellina. In the eleventh book of *De Magnificentia* and in the sixth book of his *De bello Neapolitano*, Pontano recorded how Alfonso I had enlarged the *Crypta Neapolitana*, and described the cliff and the caverns of the Platamone, which he later transfigured into the poetical image of a nymph in the *Lepidina*.[25] Sannazaro referred to the Platamone in the fifth *Egloga piscatoria*, and writes that its caverns had in antiquity been the site of a cult of Serapis.[26]

---

21   For the contacts between Leone and the Accademia Pontaniana, see the Introduction to the present volume, 1–3.

22   See Miletti in this volume, 29–41.

23   Leone 1514, *Praefatio*, f. ii *recto* (see below, Appendix 2, § 3, 165).

24   For the *Crypta Neapolitana* and the church of Santa Maria di Piedigrotta, see D'Ovidio 2006–2007; D'Ovidio 2012; D'Ovidio 2013.

25   Pontano 1498, bk. 11 (ed. Tateo) 1999, pp. 188–189: "Refertur ad Marcum Cocceium (nesciam an is fuerit Marcus Cocceius Nervae Augusti avus, qui Romae aquarum curam habuit, tulitque maximam architecturae laudem), refertur, inquam, ad hunc sive ad alium Cocceium crypta Neapolitana (sic enim ad accolis vocatur), mirum quidem opus et iter facientibus commodum. Extat altera etiam crypta, quae Seiani dicitur, qua parte Pausilypus mons in mare protenditur, cuius transitus ob ruinas impeditus est". Pontano (ed. Summonte) 1509, bk. 6, last page, folios not numbered: "Sunt geminae cryptae perforato monte Pausilypo, altera ad viam puteolanam in ipsoque fere promontorii principio, quae ab Alfonso rege fuit non modice amplificata, altera ad montis in mare prominentiam atque ad ipsius exitum eaque maxima parte ab vetustate labefacta". See Iacono 2009, 571–572; de Divitiis 2015b, 210–211.

26   Sannazaro, *Fifth Piscatory Eclogue* (5.5), ed. Putnam 2009, 134. See de Divitiis 2015b, 210–211.

That Leone was not merely aware of Neapolitan humanism but an active participant in its debates is demonstrated by the care he takes to disprove the traditional story that ancient Nola had refused to supply water to the poet Virgil for his farm.[27] Such a tradition was the subject of much contemporary controversy among the humanists of the *Accademia Pontaniana*: Pontano regarded it as untrue and attempted to refute it, while, on the other hand, Sannazaro, who also appears to have suffered some injustice on the part of the Nolans, emphasizes the story of Virgil and the town's water supply, by calling it a place which was unfriendly to poets.[28] The importance of Sannazaro for Leone is proved also by the presence in the map of the *Ager* of the no longer surviving towns of Ercolano and Pompei [fig. 1], which the Neapolitan poet mentions in his *Arcadia*. In particular, Pompei is located by Leone near the site of modern Scafati, along the river Sarno, exactly where Sannazaro had imagined it.[29]

2   The Ancient City: The *Nola Vetus*

The close connection between past and present which lies in the background of the map of the *Ager* is the main subject of the second plate, that of the *Nola Vetus* [fig. 2], representing the ancient visual appearance of the town in the Roman period. The exceptional nature of this engraving was noted several decades ago by Roberto Weiss, who underlined that the *Nola Vetus*, printed by Leone with the book in 1514, is the earliest graphic reconstruction of a no longer surviving ancient town executed in the early modern period. It predates not only the reconstructions of ancient Rome published in 1527 by Fabio Calvo,[30] but also the first and never realised attempt by Raphael to draw such a map, a project documented in the famous letter to Pope Leo X.[31] Ambrogio Leone was certainly well known in the Roman milieu of those years, since he dedicated his *Castigationem adversum Averroem* to Pope Leo X.[32] It can therefore be inferred that the *De Nola*, and with it the map of *Nola Vetus* were known

---

27   On the question of Virgil and the water of Nola, see Miletti 2016a and de Divitiis, Lenzo in this volume, 56–57.

28   See de Divitiis, Lenzo in this volume, 56–57.

29   Sannazaro, *Arcadia*, prose XII, 32 (ed. Vecce 2013, 298–299): Pompei "irrigata da le onde del freddissimo Sarno".

30   Calvo 1527; see Pagliara 1977; Jacks 1990.

31   Golzio 1936, 113; Pagliara 1977, 71; Frommel, Ray, Tafuri (1984) 2002, 396–399, 437–450; Rowland 1994.

32   Leone 1517.

to humanist circles in Rome. Besides, Fabio Calvo could also have heard of this map from Erasmus, who was a mutual friend of both men.[33]

Although it is possible that Leone had connections with humanistic circles in Rome the similarities between his *Nola Vetus* and the map of ancient Rome planned by Raphael can mainly be attributed to the fact that both take as their common model Leon Battista Alberti. In the *Ludi mathematici* Alberti explained how to make precise measurements in drawing the exact plan of a city, and in the *Descriptio urbis Romae* he gives the measurements which would allow the perimeter of the Aurelian wall of Rome to be drawn. On several occasions throughout his works Leone demonstrates a profound understanding of Alberti's theories.[34] In the description of the typical Nolan house he copied from Alberti's *De re aedificatoria* the false etymology of the Italian word *sala*, and he clearly had in mind the distinction drawn there between castle (*Arx*) and palace (*palatium*) when describing the castle of Nola and the Orsini palace.[35] He knew not only the *De re aedificatoria*, the *Ludi mathematici* and the *Descriptio urbis Romae*, but also the *Momus*, to whose protagonist, i.e. Momus, the personification of blame, Leone dedicated an entry in the *Opus quaestionum* of 1523.[36] Leone probably developed such an interest in Alberti's thought during his association with the members of the *Accademia Pontaniana* in Naples, which at the turn of the fifteenth century was perhaps one of the best places in Italy to study Alberti's legacy and become familiar with it. Jacopo Sannazaro studied the *De re aedificatoria* and wrote an index to its remarkable passages,[37] while Pietro Summonte explicitly referred to Alberti when he condemned as a "great heresy in architecture" the structural feature of arches resting on columns.[38] Even before this, the king's secretary Antonello Petrucci († 1487) possessed a codex containing the third book of the *De Familia* and built in his own palace a copy of the Albertian Ionic portal which had

---

33 For Erasmus and Calvo, see Jacks 1990, 454, note 3. For Erasmus and Leone, see Defilippis 1991; Vecce 2000; Miletti 2016a.

34 For Leone and Alberti, see Defilippis 1991. A sketch-like summary is also in Borsi 2006, 295–306.

35 Alberti (ed. Orlandi) 1966, bk. IV, ch. 5, 156–157. See de Divitiis in this volume, 97–98.

36 Leone 1523, *Problema* 404: "Qur a Graecis excogitatus deus est, quem Momum appellant, qui cuncta deridet, reprehendit, carpit". As underlined by Paola Zambelli, the interest in Momus can also be found in the *Adagia* by Erasmus (*Adagium* 5074), as a direct consequence of the appearance of the printed edition of Alberti's *Momus* in 1520; Zambelli 2004, 192–204; Zambelli 2007, 254–264.

37 Biermann 1990; Vecce 1997; Borsi 2006, 267–277; Vecce 2007.

38 Nicolini 1925, 171; Pane 1975–1977, I, 63–71 (69).

been built some years before at the entrance of Diomede Carafa's palace.[39] A manuscript of the *De re aedificatoria* was owned by the cardinal Giovanni d'Aragona,[40] whose Albertian culture can still be detected in the remains of his villa, built before 1485, where the arches of the loggia rest on pillars, rather than on columns.[41] That Leone was aware of Giovanni d'Aragona's artistic patronage is shown by the lengthy description he dedicated in his *De nobilitate rerum* to the precious embossed inkwell made for the Cardinal by Caradosso [fig. 58, 59, 60, 61], in which he also shows that he knew how much it had cost.[42]

We know that Raphael wanted to draw an exact orthogonal plan of Rome, strictly in adherence with the method explained by Leon Battista Alberti in his *Descriptio urbis Romae*, while the maps published by Fabio Calvo are pictorial bird's-eye views.[43] The *Nola Vetus* is a hybrid of the two kinds of representations. Leone represents ancient Nola on a diagrammatic map where one can see at the same time both the ancient and the modern city. He also adopted various methods of representation: if at first glance this choice might seem naive, in reality it fits perfectly with the need for graphic differentiation in visualizing the multiple levels of certitude in archaeological knowledge. The recourse to two-dimensional plans and to three-dimensional views in the same map reflects what Leone could reconstruct without doubt, what he thought to be probable and what he was unable to decode. The great number of captions which appear all across the map would seem to confirm that Leone took the leading role in the creation of this plate.

The outline of the city, i.e. the wall, is drawn as a double circle interrupted by twelve gates unconvincingly represented in elevation in a kind of

---

39   Petrucci's manuscript is now in the Biblioteca Casanatense of Rome; see Bertolini 2005, 66, 68; Bertolini 2006, 31 note 50; Borsi 2006, 108. For the ionic portal, see de Divitiis 2007, 65–77.

40   In 1483 Giovanni d'Aragona asked Francesco Gaddi in Florence for a copy of Alberti's architectural treatise. The manuscript was created for the cardinal the following year in Ferrara. See de Marinis 1952, I, 88; Haffner 1997, 94; Borsi 2006, 258–267.

41   For the villa of the Cardinal of Aragon, later known as villa Conigliera or villa Luperano, see: Lettieri [1560] 1803, 402; Pane 1937, 30–31; Pane 1975–1977, II, 58–62; Quinterio 1996, 432–433; Fiengo, Guerriero 2010. The courtyard of the royal villa of Poggioreale, begun some years later by Giuliano da Maiano, had arches on round columns. For Poggioreale, see Quinterio 1996, 438–496; Maffei 1996; Lenza 2004; Lenzo 2006, 270–271; Modesti 2014; de Divitiis 2015; Lenzo 2015.

42   Leone 1525, cap. 41, folios not numbered (see below, Appendix 5). See Morelli 1800, 205–206; Brown, Hickson 1997, 20–24. For a full discussion see Loffredo in this volume, 111–115.

43   Pagliara 1977; Jacks 1990; Carpo 1998; Carpo 2005.

unconventional 'radial' projection from the centre. Seven of these gates lead to the main roads linking Nola to the nearby towns, whose names are inscribed directly on the map. As well as the urban gates, the main buildings too are depicted in elevation, but in a three-dimensional view reinforced by the shading. In the text Leone reconstructs in detail the ancient appearance of the two amphitheatres, giving also their diameter and using them as a basis for conjectures on the overall size of the city. Therefore the more detailed outline of the two amphitheatres, together with the funerary mausoleums outside the city, the temples of Augustus, of Mercury, of Jupiter and of *Victoria*, on the map indicates Leone's greater degree of certitude that he had identified correctly their function and their location [fig. 9, 11, 12]. On the other hand, the remains of a great wall (*paries anti*[qua]) [fig. 9], which Leone interprets as belonging to an unidentified ancient building, is depicted only in plan, and as the vestiges were visible on the ground, without any attempt to reconstruct the wall's original context.[44] Antiquities found during archaeological excavations are even less well defined, and on the map they are indicated only by legends reading "here antiquities have been excavated" and "throughout this area antiquities have been found when men were digging" [fig. 11, 12].[45] The large square hole in the soil next to the inscription *antiqua excava*[ta] is perhaps the earliest depiction of a modern archaeological excavation. The legends added to the map are also used for Leone's conjectures about the possible presence of ancient remains still buried under the soil, conjectures he based on his perception of the way the ground surface swelled in certain places.[46]

The only modern building depicted in three-dimensional view is the round tower located at the south-west corner of the modern city wall [fig. 2, 9]. The tower has a prominent position, occupying the exact geometrical centre of the ancient circular city. Such a coincidence suggests that the tower was used by Leone in order to make the observations and measurements which were necessary for him to draw the map of the city, thus following the method recommended by Leon Battista Alberti. In the *Ludi mathematici* Alberti explained how to map a city after measuring it from a central and elevated position, and in the *Descriptio Urbis Romae* he chose the Capitoline hill for this purpose.[47]

---

44   Leone 1514, bk. I, ch. 8, f. xiii *verso*; bk. II, ch. 1, f. xxii *recto*.

45   The inscriptions are "Hic excavantur antiquitates" and "Per haec loca cava[n]tibus inventae s[unt] antiquitates".

46   Next to the southern gate is written "Hic extumet solum ob fabricam: videtur fuis[se] porta cum turr[ibus] et murorum pars" and a similar caption accompanies the image of the gate toward Sarno: "Item hic extumet solum: vide[tur] fuis[se] porta cum turr[ibus]".

47   Carpo 1998; Carpo 2005.

It seems likely that Leone followed the same method and that in order to make the necessary measurements for the map of *Nola Vetus* he climbed to the top of the round tower.

The form and the exact identification of the ancient wall was only Leone's inference, and as a consequence he represents it in a different way : it is more sketchily indicated and accompanied by several captions specifying where the rise in the surface of the soil suggested the existence of buried buildings such as urban gates framed by twin towers. In order to make it clear that it was not certain the wall was circular, he also drew on the map, as an alternative hypothesis, three sides of a possible polygonal perimeter.[48] Faced with the problem of having to envisage the perimeter of the ancient city, Leone tended to use familiar models. He could have known several examples representing circular and polygonal cities, from late antique manuscripts such as *Gromatici Veteres* and the Vatican Virgil to those found in printed books such as the *Hypnerotomachia Poliphili* and Fra Giocondo's edition of Vitruvius.[49] He might have profited from the advice of one of the skilled cartographers who were working in those very years in Venice, such as his compatriot Bernardino Silvano, or the Florentine Francesco Roselli, who was present, together with Leone, Fra Giocondo and many others at the lecture given by Luca Pacioli in the church of San Bartolomeo of Rialto, Venice on the 11th August 1508.[50] However it is also the case that Leone seems to have retained his memory of the example of the Aragonese parchment maps of the Kingdom of Naples, since he inserted in the plate of *Nola Vetus*—as in that of the *Ager Nolanus*—not only

---

48   Leone 1514, bk. I, ch. 8, ff. xi *recto–verso*.

49   For the circular maps of Rome, see Frutaz 1962; Scaglia 1974; Cantatore 2005. On round cities in general, see Moffitt 1993. For the *Vatican Virgil* and the *Gromaticos Veteres* see Nolhac 1897; Martines 1976; Martines 1976a; *Fragmenta et pictura Vergiliana* 1945. For late medieval and early modern two-dimensional plans of cities, such as the anonymous maps of Talamone (1306) near Siena, Mali Ston (1396) in Dalmatia, Verona (1453–59), and those of Ferrara (1494–95) by Prisciani, Imola (1502) by Leonardo, and Pisa (1508–13) by Giuliano da Sangallo, see Almagià 1925; Schulz 1978; Friedmann 2010; Friedmann 2010a; Folin 2010; Folin 2010a; Nuti 2010; Nuti 2016; Santucci 2017.

50   Euclid (ed. Pacioli) 1509, f. 31 *recto*; Lepori 1981; Danzi 2005, 25–26, note 38. "Frater Iocundus Veronensis Antiquarius. Omnes prelibati Eiusdem Minoritanae Familiae" is recorded among the "Reverendi Sacre Theologie Professores". The list includes also Marin Sanudo, Giovanni Bembo, Aldo Manuzio, the mapmaker Francesco Rossello, an unknown "Joannes Franciscus puteolanus" and the architects "Georgius Tragurinus eiusq. filius Marcus, Alexius Bergomensis, Ioannes Marcus Canotius Patavinus, Petrus Lombardus". For a recent discussion of the subject, see Benzoni 2014. I am grateful to Francesca Salatin for indicating to me this study.

geographical and chorographical but also historical information. For instance, Leone indicated on the map the site of Hannibal's military camp outside the wall. He had no evidence for its exact location, but he required this information to be shown on the map, as he wished to stress the loyalty of the Nolan people to their allies and friends when they refused to enter into a treaty with Hannibal (Livy, 23.44), an attitude which was in stark contrast to the cowardice displayed by the nearby rival city of Capua.[51]

3   Comparing the Ancient City and the Modern One: The *Figura Praesentis Urbis Nolae*

The efforts by Leone to create a precise reconstruction of ancient Nola are apparent in the third plate, representing the *Figura praesentis urbis Nolae* [fig. 3]. This is the only one of the four plates which is not a map. It shows a series of geometrical plans arranged in three horizontal rows. The first shows the perimeter of modern Nola, which in the text Leone describes as being in the shape of half a footprint.[52] The second row is occupied by four smaller regular geometrical figures, while in the bottommost one there is a double circle and a square subdivided into other squares and rectangles containing letters and numbers; beneath the rows, at the bottom of the plate, three small plants are depicted.

The purpose of this strange set of geometrical figures is to help the reader in understanding what Leone writes in book II, chapter 2. Here he attempts to demonstrate, using seven arguments, based on geometry and an algebraic matrix, that ancient Nola was much larger than the modern city. First Leone explains that the city has eleven corners and is in the form of half a man's footprint. Then he gives the dimensions of the perimeter of the modern city, which measures 924 steps, and calculates the extent of the area it encloses. After this, he demonstrates with five examples that the form of a geometrical figure is unimportant, since what matters is the area. Thus he can assume that the dimensions of modern Nola are identical to that of a circle with the same area, and finally he can compare such a circle with another one which has the dimensions of the ancient city. The conclusion of his argument is that the ancient city was six times larger than the modern city of Nola. And, if further

---

51   Leone 1514, bk. ch. I, 13, ff. xviii *recto–verso*. For the Renaissance interpretation of the episode of Capua's betrayal of Rome, see Miletti 2014.

52   Leone 1514, bk. II, ch. 2, f. xxii *verso*.

confirmation is needed for this conclusion, he provides an explanation of how to draw a complicated matrix which uses algebraic computations to prove it.[53]

The only difference between the geometrical explanations given by Leone in the text and demonstrated in the plate is that in the plate the fourth and the fifth stages in Leone's calculation are merged into a single geometrical figure composed of a hexagon and a rectangle attached together. This was probably an error on the part of Mocetto, who may not have completely understood the drawing which Leone gave him. The two circles in the third row show the dimensional comparison between the area of the modern city and the ancient one, while the square on the right is a mathematical matrix demonstrating the same proportional relation not with geometrical models but through algebraic computations.

It is clear that this complex plate must have been drawn by Leone, and that probably the only addition by Mocetto were the three little plants at the bottom of the sheet.[54] This plate, together with the corresponding chapter of the *De Nola*, demonstrates that Leone was proficient not only in Greek and Latin humanities, but also in mathematics. Once in Venice, he could have profited from his acquaintance with Luca Pacioli, but he must have had his first mathematical training in his native city. That mathematical culture was widespread in Nola is demonstrated by Pietro Muscarello's *Algorismus* [fig. 39, 40], a manuscript treatise on algebra, dated to 1478 and composed in Leone's native city for the Albertini family.[55] Leone certainly knew Muscarello, since in the *De Nola* he mentions him in relation to the theatre performances he arranged together with his father Giovanni.[56]

The presence of sophisticated mathematical arguments in the text of *De Nola* is not as striking as Leone's choice to devote one of the four plates of the book to this aspect of his subject. He wanted to underline the close connection between ancient and modern Nola in order to communicate in a more effective way, even with the use of geometrical diagrams, the continuity of the city's magnificence from antiquity down to his own day. And even more importantly, Leone's contemporary readers needed to visualise the fact that ancient

53   Leone 1514, bk. II, ch. 2, ff. xxii *verso*–xxiii *verso*.

54   Hind 1938–1948, part II, V, 171 wrote of a copy of the engravings without the three flowering plants added along the lower margin (London, British Museum Library, C.178.g.14), which I have not been able to examine. I thank Stephen Parkin for having verified this information for me and confirmed that the book is now in the British Library, shelfmark 178.G.14.

55   Muscarello (ed. Chiarini) 1972. The manuscript is now in Philadelphia, Penn Library, *Lawrence J. Schoenberg Collection*, LJS 27. See also de Divitiis in this volume, 82.

56   Leone 1514, bk. III, ch. 10, f. liiii *recto*.

Nola was at least six times larger and more monumental than the city they could see with their own eyes or infer from the merely verbal descriptions in Leone's text.

The *Figura praesentis urbis Nola*, with its complex and subtle mathematical demonstrations, could only have been conceived and drawn by a scientist such as Leone, rather than a painter like Mocetto. Such a drawing was executed on the basis of measured plans of Nola's walls that Leone must have brought with him to Venice, but which he clearly sketched much earlier. It is worth noting that in all four plates in the book the map of modern Nola is represented. It appears in the *Ager Nolanus* [fig. 1, 8], it is depicted inside the great circular shape of *Nola Vetus* [fig. 2, 9], it is the object of the mathematical demonstrations in the *Figura praesentis urbis Nolae* [fig. 3, 10], and finally it is the main subject of the fourth and last map [fig. 4]. Leone stresses the continuity between the ancient and the modern town, as he wanted to emphasize how the ancient and glorious history of Nola could be seen reflected in the modern city. The reliability of the architectural and urban surveys of Nola at Leone's disposal is further demonstrated by the plate representing *Nola Praesens*.

4   The Glory of the Modern City: The *Nola Praesens*

The last engraving of the book is devoted to modern Nola, shown in a bird's-eye view as was usual in urban depictions of this time [fig. 4]. The most striking feature of the city is the double wall, a lower external wall and a higher internal one, both divided by towers at frequent intervals. The names of the four gates and of the main towers are engraved on the plate and correspond to the description given by Leone in chapter 7 of book II.[57] A moat encloses the exterior wall, and the road alongside it is guarded by knights and foot soldiers carrying spears. Close to the wall is the square castle (*Arx*) [fig. 36], with round towers at each corner, plus a square tower in the centre of the southern façade, and a larger round one in the middle of the court.[58] The comparison with later plans and views [fig. 37, 38] demonstrates that the castle, and most probably the entire fortification system, was depicted starting from detailed architectural plans and elevations containing measurements, later transformed into a perspective view.[59] Outside the gate named *Portellum*, on the west side (*occasus*), there are some trees and a fountain: this must be the *forum boarium*, or cattle

---

57   Leone 1514, bk. II, ch. 7, f. xxviii *verso*.

58   Leone 1514, bk. II, ch. 8, ff. xxix *recto–verso*.

59   Carillo 1996; Caianiello 2003.

market, where count Orso Orsini had had several rows of linden trees planted to provide shade from the sun for the merchants as they bargained.[60]

Inside the wall there is no indication of ordinary private houses: only the main buildings are depicted, such as the cathedral (*episcopium*), the other churches, the covered market (*dogana*), the seggio (*porticus*) [fig. 33], the monastery of San Francesco and the convent of the Rocchettine (*collegium*) and the palace of the count (*Regia*) [fig. 28]. The centre of the city is occupied by the great square (*area*), opened up in front of the cathedral between the streets Vicanciana, Portellana, Cortefellana and an unnamed road linking the last two. In the second book of the *De Nola*, Leone explains that this square was the *forum frumentarium*, and that it had been enlarged and restored by count Orso Orsini.[61] The cathedral is clearly depicted with three aisles and a high transept. The bell tower rises alongside on a base of massive stone blocks in *opus isodomum*,[62] while at either end of its transept can be seen the little churches of St. John the Baptist (*D. Io. B*[a]*p*[tista]) and the Holy Apostles (*S. Apostolus*). Overlooking the square there is the seggio, the building where the nobles met, and the church of St. Felix. The seggio appears as a low building rectangular in plan. The map shows two of its sides, the shorter of which is completely blind, while one half of the longer side is opened by an arch.[63] The church of St. Felix is depicted as a small building, since its lower section was underground.[64] Behind it, facing the via Portellana, there is the *Dogana*, a covered loggia opened by three arches, which, since it had fallen into a state of

---

60   Leone 1514, bk. II, ch. 7, f. xxix *recto*.

61   Leone 1514, bk. II, ch. 16, f. xxxx *verso*. In the square of the cathedral there was a large marble slab, resting on a ancient marble base, on which all the standard local weights and measures were inscribed. It had a chain attached at one end and thus also served as a public pillory for the punishment of minor crimes; see Leone 1514, bk. II, ch. 17, f. xxxx *verso*.

62   Leone 1514, bk. II, ch. 11, ff. xxxi *recto*–xxxiii *verso*, gives a very detailed description of the cathedral and the adjoining buildings; on this subject, see the chapter by de Divitiis.

63   Leone 1514, bk. II, ch. 10, f. xxx *recto*, describes the seggio as a rectangular building, five steps long and 20 feet in width. It was 10 feet high. It had a loggia opened by two arches resting on a single corner column 10 feet high; the lower sections of the arches were closed by low walls. The interior was covered by a vault and divided into two parts, the loggia and a more private room, while running the length of the perimeter there was a stone bench. The entire building was raised above street level on a pedestal of six steps. For the seggio of Nola, see Lenzo 2014, 178–179, 200.

64   Leone 1514, bk. II, ch. 13, f. xxxvii *recto*.

disrepair, had been rebuilt by count Orso, and was used as a market for flour, oil, wax, leather, clothes, and other merchandise.[65]

Detailed attention has been paid to the actual position of the main buildings within the city, even when this is disadvantageous to the general view. For example, only the rear side of the count's palace (*regia*) is shown [fig. 28], while its rich façade which Leone praises on account of its re-use of ancient blocks taken the "marble amphitheatre" remains hidden.[66] The same method is adopted for the churches, whose façades are oriented towards the west (*Episcopium*, *S. Paulinus*, *S. Felix*, *S. M*[aria] *Nova*, *S. Spiritus*, *S. Servator*), southwest (*S. Clara*), north (*Sacellum Iesu*), or south (*S. Francis*[ci]), depending on their real location.[67]

While such buildings are represented in three-dimensional view, the urban structure is suggested by a two-dimensional plan of the streets, each of which is clearly identified on the map. A similar mode of representation is also found in the medieval urban iconography of Venice, as in the case of the map drawn by Elia Magadizzo for the doge Ordelaffo Falier (1102–1118), known from later copies, or the one realized by Cristoforo Sabbadino in 1557.[68] It is possible that Leone knew such maps, but it is worth noting that the view of *Nola Praesens* has many points in common also with the illustrations found in illuminated anatomical manuscripts,[69] since the city is represented as though it were a single living body, with an exterior shell enclosing the main organs of urban life, and a network of streets similar to an arterial tree. Being a physician, Leone might have been familiar with this kind of illustration.

However, it is clear that the map of *Nola Praesens* was composed by putting together several detailed drawings. When the map is compared with the corresponding passages in the text, it shows that Leone must have brought with him to Venice a large portfolio of architectural drawings which included, at least, the measured plans of the urban wall, the cathedral, the castle and the seggio,

---

65   Leone 1514, bk. II, ch. 17, ff. xxxx *recto–verso*.

66   For the Orsini palace, see de Divitiis 2016 and de Divitiis in this volume, 84–86.

67   The churches of Nola are described in Leone 1514, bk. II, ch. 13, ff. xxxvi *recto–xxxvii verso*. Between the monastery of San Francesco and the wall there is a curious building resembling a chimney, possibly a bakery or a kiln, not identified on the map or mentioned in the description of the monastery.

68   Only three copies of Magadizzo's map survive: the best known is the one in Venice, Biblioteca Nazionale Marciana, cod. lat. Z-399, n. 1610, fol. 7r, which was first described and reproduced in Temanza 1781; see Schulz 1978, 445. For the map by Sabbadino, see Svalduz 2010.

69   On scientific illuminated manuscripts, see Weitzmann 1959, 5–30.

and perhaps also plans and elevations of other buildings, such as the Orsini palace and the church and monastery of San Francesco. The illustrations for the *De Nola* were planned and drawn on the basis of this set of drawings.

5   Leone and Mocetto: Problems of Method and Authorship

There is little doubt that Leone had on his writing desk a set of architectural drawings and sketches when he began to compose the *De Nola*. He must firstly have used this graphic material as a resource for his verbal description of the main ancient and modern buildings in Nola. When he decided to add illustrations to his text, these drawings proved to be even more valuable, but he used them in a different way. If in the text all the drawings are precisely described, only in the *Figura Praesentis urbis Nolae* does the plate give a direct representation of what is described in words. This is not the case for the other plates. There is neither a plan of the cathedral, nor of the seggio, although in the text Leone also gives the measurements of the buildings.[70] And there is no representation either of the typical Nolan house, the detailed discussion of which, also in this case including precise measurements, reveals Leone's knowledge also of the theoretical problems of architecture.[71] The four engravings are different. The map of the *Ager* serves to show the complex network of relationships between Nola, the surrounding countryside and the nearby towns; that of *Nola Vetus* demonstrates the continuity of the city from antiquity to the present day on the same site; thus the reduced size of modern Nola when compared with the ancient one, as illustrated on the *Figura praesentis urbis*, appears as a direct consequence of this persistence over many centuries of history. Finally, the map of *Nola Praesens* shows the spatial and visual relationships of the different parts of the city both between themselves and with the whole, conveying through this Albertian *concinnitas* the image of an almost ideal city, of a perfect society living in harmony. The relationship between the text and the images in the *De Nola* thus has a double direction: on one hand written descriptions enable readers to form in their minds the images conceived by the author; on the other the plates allow the many concepts analytically expressed by the words to be communicated and grasped in a rapid synthesis, thus offering another and more immediate level of reading. Studies of the volume have hitherto considered the plates and above all the three maps mainly as a graphic resource for the reconstruction of the appearance of Nola in the ancient and

---

70   Leone 1514, bk. II, ch. 10, f. xxx *recto*, for the seggio, and bk. II, ch. 11, ff. xxxi *recto–xxxiii verso*, for the cathedral.

71   On Leone's discussion of the Nolan *domus*, see de Divitiis in this volume, 92–100.

early modern period. However, as we have seen, if the text of the book can be seen as self-contained, the images, on the contrary, can be understood only if analysed together with the text. It is probable that it was only after Leone had arrived in Venice that he fully understood the great communicative potential of illustrated books such as the *Hypnerotomachia Poliphili* and the new edition of Vitruvius' *Architectura* published by Fra Giocondo: the excitement of this discovery can be glimpsed in the *Praefatio* of the *De Nola* when Leone praises the "almost divine strength" of images.[72]

Although it is certain that Mocetto engraved the plates, it is difficult to distinguish the parts he and Leone played in making the preliminary drawings. Most probably they worked on them side by side. Mocetto's signature on the plate of the *Ager Nolanus* [fig. 5, 6] might suggest that he thought this plate to be more his own creation than the others. But, on the other hand, the importance there attributed to historical information, the exact knowledge of the places, and the perfect agreement with the text—both in detail as well as more generally—make it clear that Leone had a predominant role in the composition of this map of the territory.

In the map of *Nola Vetus* [fig. 2] the use of plans and isometric views on the same plate might suggest that Leone and Mocetto played distinct roles, though the hypothesis should be treated with caution. Leone could have decided the general composition and given a basic design to Mocetto, who reworked it and engraved the plate. In book I, chapter 8, Leone explains how he was able to confirm that the two amphitheatres in Nola had a diameter of 50 feet and then, extrapolating from the extent of their foundations, he conjectured that the "marble amphitheatre" had three levels of arcading, while the "brick amphitheatre" had only two.[73] Since he specifies the dimensions of the two amphitheatres, and explains how he reconstructed their plans and elevations, we may again assume he had with him architectural drawings. In the map the amphitheatres are drawn in an exact orthogonal elevation in part disguised by strong shading [fig. 11, 12], thus departing from the bird's-eye view frequently found in Renaissance reconstructions [fig. 18] which derived from ancient Roman representations cast on coins or carved in stone [fig. 13].[74] However, even if we know that Leone collected ancient coins,[75] it seems that he chose

---

72   Leone 1514, *Praefatio*, f. ii *verso* (see below, Appendix 2, § 6).

73   Leone 1514, bk. I, ch. 8, f. xi *recto*.

74   This was the case of Fabio Calvo and Pirro Ligorio; see Pagliara 1977; Burns 1988; Jacks 1990. Similarly, in the Aragonese parchment maps, the amphitheatre of Nola is shown from a bird's-eye point of view; see Cesarano 2012.

75   Marcantonio Michiel states that an ancient silver coin from Syracuse, with a crowned female head surrounded by dolphins, listed in 1530 in the Venetian collection of Marco

to ignore such a source. He based his archaeological reconstruction on the direct knowledge of the ancient architectural monuments he had seen during his travels throughout Italy.[76] It is therefore clear that the preliminary drawings of the amphitheatres, being architectural drawings, owe more to Leone than to Mocetto.

The same is probably true also for the other buildings depicted on the map of *Nola Vetus*, which instead reveal their derivation from a scholarly antiquarian tradition. As Leone writes, archaeological evidence was scarce. Thus, in the plate of *Nola Vetus* the images of temples, baths, city gates, and funerary monuments serve mainly to indicate their existence and location, appearing like pieces on a chessboard, whose only function is to fix a position. While in the text it was enough to provide a simple explanation of what kind of building they had been, and where their remains were located, on the map it was necessary to give them a form. Some years later, Fabio Calvo, when he had to draw the ancient temples of Rome, referred to the miniatures of the Vatican Virgil.[77] Thus in his *Simulachrum urbis*, Roman temples are depicted as rectangular buildings with columned façades and constructed on high podiums. Leone might have known the Vatican Virgil or similar manuscripts; besides, columned rectangular temples still existed in Southern Italy, such as the temple of Castor and Pollux in Naples and that of Augustus in Pozzuoli; moreover, the Greek temples of Paestum were known to humanists, as shown by Pietro Summonte's letter to Marcantonio Michiel in which he describes them as *opera dorica*.[78] Nevertheless, in the *Nola Vetus* by Leone and Mocetto the temples are shown as central buildings covered by domes. Many ancient domed buildings in the neighbourhood of Naples were at this time identified as temples. In Pozzuoli some remains of Roman baths were identified as temples of the Sybil, of Diana, of Apollo, and of Neptune [fig. 16], and in Naples the ancient structure within which the church of Santa Maria Rotonda stood was believed to have been the temple of Vesta.[79] Also in the Venetian *Hypnerotomachia Poliphili* the

    Antonio Foscarini, had previously belonged to Ambrogio Leone; see Morelli 1800, 69, 204–207. Leone himself describes another ancient coin from Gela depicting a bull with the head of a man; see Leone 1523, n. 403.

76   Leone cites the amphitheatre of Capua, and he could also have known those of Pozzuoli, Roma, Minturno, Verona and Pola.

77   Pagliara 1977; Jacks 1990.

78   On the temple of Naples, see Lenzo 2011. On that of Pozzuoli, see Dubois 1907, 345–347; Maiuri (1934) 1958, 34–35; Adinolfi 1968; D'Ambrosio, Giamminelli 2000; Valeri 2005. On the letter by Summonte to Michiel, see Nicolini 1925; Pane 1975–1977, II, 63–71; Bologna 1995.

79   For the ancient round buildings of Pozzuoli see Dubois 1907; Maiuri (1934) 1958. For the church of Santa Maria Rotunda in Naples, see Lenzo 2011, 48, 69 note 277. Another circular

ancient temple of Venus is shown as a round building.[80] At the beginning of the sixteenth century the most magnificent ancient temple in the eyes of humanists was without doubt the Pantheon in Rome, so it is unsurprising that Leone thought of the temples of *Nola Vetus* as centrally planned domed buildings. Among them the temple of Mercury [fig. 14] seems to recall faintly the church of San Francesco in Rimini as shown in the foundation medal cast by Matteo de Pasti [fig. 15]. A possible derivation from the goldmith's art might instead be conjectured for the unconvincing reconstruction of the two funerary mausoleums (*antiqui tumuli*), shown outside the wall of the ancient city [fig. 20]. When compared with existing remains and with drawings dating from the mid-sixteenth century [fig. 21, 22], the reconstructions shown in the *De Nola* are overly disproportioned. The main lower structures appear to be compressed by the heavy roofs, depicted as improbable onion domes, which, rather than being based on archaeological evidence, seem inspired by small metalwork artefacts such as Byzantine reliquaries and censers that must have been very familiar in Venice, or even inkwells such as the one made by Caradosso for cardinal Giovanni of Aragona.[81]

As for the plate of *Nola Praesens* [fig. 4], we can suppose that Leone gave to Mocetto architectural drawings depicting the wall and the main buildings, and that the painter assembled the several drawings into a single view, even if the incorrect perspective projection of the west part of the urban wall is inconsistent with the mastery of such a technique which one can normally attribute to this artist. The bird's-eye view of *Nola Praesens* befits the technical knowledge possessed by a painter better than the plan of *Nola Vetus* and the geometrical figures of the *Figura praesentis urbis Nolae*. But the different kinds of representation employed might also correspond to a profound understanding of Leon Battista Alberti's thought. Alberti did not condemn the technique of perspective, merely stating that architects should not employ it but leave it to painters to use. According to Giorgio Vasari, Alberti himself might have painted a urban bird's-eye view of the city of Venice.[82] Therefore, the use or otherwise of perspective depended on the purpose for which it was intended. Alberti painted

---

ancient temple dedicated to Minerva is shown at Punta Campanella, near Sorrento, in the eighteenth-century copy of a lost Aragonese parchment map; see La Greca, Valerio 2008.

80 Colonna 1499, n3r.

81 A relevant example might be the silver reliquary built in the form of a domed church which is still preserved in the Tesoro of San Marco in Venice; see Gaborit-Chopin 1984. For the hypothetical reconstruction of Caradosso's inkwell by Brown, Hickson 1997, see in this volume fig. 58.

82 Vasari 1550, 376: "figurò ancora una Vinegia in prospettiva, e San Marco: ma le figure che vi sono furono condotte da altri maestri. Et è questa una de le migliori cose che si vegga di suo di pittura". See Bulgarelli 2002.

Venice in perspective because the aim was to have an effective image of the existing city, while he drew Rome in orthogonal plan because his purpose was to measure the perimeter of the walls. The same difference existed in the *De Nola*. The modern town is portrayed in perspective like Venice, while the ancient one is reconstructed as accurately as possible, and thus drawn in orthogonal plan like the Rome of Alberti.

A proof that Mocetto did not work independently even when engraving the copper plates comes from the evidence of some last-minute changes of detail, after the plates had been finished, which it is certain were required by Leone. On the map of *Nola Vetus* one can still see the traces of correction on the depiction of the thermal building (*Antiquae balneae*) [fig. 11, 12], which at first was engraved with its roof turned eastwards and floor towards the west, and subsequently was rotated ninety degrees counter clockwise. In the *Figura praesentis urbis* [fig. 3] the only mistake, which was not corrected, is the merging into a single figure of the fourth and fifth stages of the geometrical demonstration, respectively a rectangle composed of two and a half squares and a hexagon. The *Nola Praesens* [fig. 4] is technically of higher quality, and the only revision which can be detected is in the *Turris Prioris*, one of the many towers dividing the stretch of urban wall, which was re-engraved lower than it had previously been shown.

By adding the four engravings to the text of the *De Nola*, Leone strongly enhanced the impact of his book. The plates helped to fulfill the author's main aim, which was to transmit an image of Nola to posterity and complete the information given in the text. At the same time the insertion of images in a chorographic work emphasised the pioneering character of the *De Nola*, marking a shift in the antiquarian and topographical genre for which some kind of graphic component would become an almost indispensable feature.[83] The importance of Mocetto's contribution to the illustrative apparatus of the *De Nola* cannot be underestimated, and is clearly underlined by Leone in his introduction. However, the many questions which surround the four engravings suggest that the Venetian painter acted as little more than an executant. The stratified layers of knowledge which can be read in the engravings, such as the simultaneous connections with geography and history, the precise and accurate familiarity with places and buildings, and the existence and nature of the corrections subsequently made to the finished plates would all seem to suggest that Leone should be regarded as their real author.

---

83   Weiss 1958, 178–179.

CHAPTER 4

Architecture and Nobility: The Descriptions of Buildings in the *De Nola*

Bianca de Divitiis

Among fifteenth and sixteenth-century chorographical works the *De Nola* stands out for the notable attention Leone pays to architecture. Descriptions of buildings form one of the main narrative cores of the work where, by the means of a sophisticated terminology, the humanist reveals a profound knowledge both of building sites, materials and techniques, and specialized literary works, such as the treatises of Vitruvius and Alberti.

By analyzing the nature and the form of the lengthy digressions on the main buildings of the city and in particular the exceptional chapter devoted to the typical house which could be found in Nola, this essay will focus on Leone's familiarity with the architectural world, both in terms of theoretical and practical knowledge. It will also examine the role attributed by Leone to architecture, as one of the most tangible forms of expression of his idea of *nobilitas*.

The close relation between architecture and nobility that Leone expresses in the *De Nola* is grounded on the major contemporary humanistic debates, as that on the idea of nobility, the *paragone* among the arts, and the more specific discussions on the theme of the house. In this context the role of architecture as explicitly described in the *De nobilitate rerum dialogus*, the treatise on nobility which he composed approximately fifteen years earlier, will be regarded at the same time as a precedent and a theoretical reflection on ideas that will find in the exceptional descriptions of the *De Nola* a full literary accomplishment.[1]

1 Leone and Architecture

Leone was a doctor and not a professional architect. Certainly though, by the time he wrote the *De Nola* he had acquired all the multiple architectural skills which allowed him, if not to design buildings, at least to describe them.

---

[1] Leone 1525. For a general description and dating of the *De nobilitate* see de Divitiis, Miletti 2016.

© KONINKLIJKE BRILL NV, LEIDEN, 2018 | DOI:10.1163/9789004375789_006

Such competence had been built up over time, from his earliest years in Nola, where he was educated in an environment imbued with geometric and mathematical culture, and where building was a prominent activity in the city and in his family. Leone recounts that his uncle was involved in the construction of the Nolan cathedral as director of works (*fabricae praefectus*), as well as being responsible for all the architectural and, in general, business affairs of the Orsini family.[2] Furthermore, in the *De Nola* he mentions several figures connected to the architectural world who were active in Nola at the time, such as Renzo Martino, Napolitano Fiorenzo and Barone Perusino, whom he describes as *geometra*, as well as Marino Sasso, whom he also goes as far as calling *architecturae peritissimus*.[3] These acquaintances would have granted him privileged access to the many building sites ongoing in the city and possibly also to the drawings. A sense of how architecture was an important economic activity in the city comes from the mathematical treatise entitled *Algorismus*, written in Nola in 1478 by another acquaintance of Leone, Pietro Paolo Muscarello, who among various mathematical problems cites the case of a nobleman who has to build his palace in thirty days [figs. 39, 40].[4] Nola could also boast a patron saint, the refined bishop Paulinus, who enriched his poems with architectural descriptions.[5] This local familiarity with architecture and this initial training would have enabled Leone to understand and appreciate the quality of the architecture in Nola, including both the remains of Roman monumental buildings and the works commissioned by the Orsini family, in particular by Orso Orsini, of which he had been a direct witness during his youth.

Leone's first contact with architecture through his city of origin and his own family must have been notably deepened and refined during his sojourns in Naples. Here he seems to have come into contact not only with the city's scientific and medical circles, but also with members of the Pontano academy, such as Giovanni Pontano himself and Jacopo Sannazaro, humanists who were all deeply interested in the material evidence of antiquity and in new contemporary architecture. They also, dedicated themselves to the study of the treatises of Vitruvius and Alberti and were directly involved in the architectural and

---

2 Leone 1514, bk. II, ch. 11, f. xxii *verso*.

3 Leone 1514, bk. III, ch. 5, ff. xxxvi *verso*, xxxviii *verso*. See Borsi 2006, 304.

4 Leone 1514, bk. III, ch. 10, f. liiii *recto*. From Leone we know that Muscarello was also involved in sacred representations which in the fifteenth century customarily took place in Nola. See Muscarello (ed. Chiarini) 1972.

5 On Paulinus and architecture see Lehmann 1998, 93–104; Goldschmidt 1940.

urban projects sponsored by the Aragonese royal family.[6] Leone's direct knowledge of the magnificent works which were being carried out in Naples at the time is proved by the detailed description of the Gran Sala in the Castel Nuovo in the *De nobilitate* and by the praise of the new walls built by Alfonso, Duke of Calabria, the future Alfonso II, in the author's preface to *De Nola* [figs. 5, 6].[7]

We cannot exclude the possibility that Leone had already met the antiquarian and architect Fra Giocondo da Verona while he was working in Naples for the royal Aragonese court at the end of the fifteenth century, but opportunities to get to know him must have increased in Venice, where they both lived at the beginning of the sixteenth century and were part of the philo-Hellenic circle of Aldus Manutius and of the school of Rialto.[8] We even have the striking coincidence of their presence in the same place, since both men attended the inaugural lecture that Luca Pacioli delivered on the fifth book of Euclid's *Elements* on 11th August 1508, in the small church of San Bartolomeo in Rialto [fig. 63].[9] In Venice Leone would have been able to see the construction of the *all'antica* project of the Fondaco dei Tedeschi and the creation of what can be considered an authentic architectural monument in paper, Fra Giocondo's new edition of Vitruvius' *De architectura*, published in 1511 [fig. 48].[10]

It was through this long course of study and contact with the most advanced humanistic ambiences imbued with antiquarian and architectural culture and through his direct experience of the major renovation works in the cities in which most of his life was passed, that Leone was able to attain a personal competence in architectural matters, form the more practical skills of measuring, inspecting building sites and identifying different materials, to purely

---

6   On the role of Pontano and of the other humanists of the Academy in contemporary architectural researches and debates see Vecce 1996 (1997); de Divitiis 2010; de Divitiis 2014; de Divitiis 2016. On Leone's contacts with Pontano and Sannazaro see in this volume Introduction, 2; Miletti, 27–32; Lenzo, 63–66; Loffredo, 106–109; Imbriani, 138–141.

7   Leone 1514, *Praefatio*, f. ii *recto* (see below, Appendix 2, § 3). Leone 1525, ch. 42, f. h iii *verso*: "Ad haec quid de arcubus dicetis, quid de testudinibus fornicibusque refractis inter sese atque per angulos prolatos earum subductis lapideis filis, qualis est Neapoli in triclinio maximo Castellinovi, in quibus mensurae rationem an ingenium efficiendi operis prius probetis ambiguum est". Cfr. de Divitiis 2016; de Divitiis, Miletti 2016.

8   On Fra Giocondo in Naples see de Divitiis 2014. On the possible contacts between Fra Giocondo and Leone see the Introduction to this volume. The most recent studies on Fra Giocondo can be found in Gros, Pagliara 2014.

9   Euclid (ed. Pacioli) 1509, f. 31 *recto*; Lepori 1981; Danzi 2005, 25–26, footnote 38; Benzoni 2014, 190–191.

10   On the projects for Rialto see Tafuri 1985, 50–61; Calabi, 1992–1993, 190–201. Molteni 2016. On Fra Giocondo's edition of Vitruvius see Giocondo 1511. Pagliara 2014; Fontana 2014.

intellectual ones, such as the study of specialised texts, drawing plans and elaborating an abstract model out of multiple built examples. All these skills lie at the origin of his descriptions of buildings in the *De Nola*.

2 The *Arx*, the *Regia* and the Seggio

The main architectural descriptions of the *De Nola* are concentrated in Book 2, where Leone tries to provide a picture of the magnificent physical appearance of the Nola of his time, namely—as he writes—in the year 1512.[11] This section comes after Leone's accurate antiquarian reconstruction of the ancient city in Book 1, where he had already stated that the remains of antique buildings offer the clearest demonstration of the nobility pertaining to the inhabitants of the city.[12]

In this context, architecture appears as one of the key elements by means of which Leone succeeds in describing the direct continuity of the noble customs of the contemporary inhabitants of Nola with those of their ancient forbears. After demonstrating that the modern city was built on the same site as the ancient one, Leone describes its main public and semi-public buildings.[13] Following an ideal path from the perimeter to the centre, he begins by describing the castle of medieval origins built over the city walls and renewed recently, about 1460, by the count Orso Orsini.[14] The detailed description of the "square island" surrounded by a moat, with four round towers at the angles and a larger tower at the centre, is complete with measurements of the entire perimeter and of the walls connecting the single parts.[15] The comparison with later surveys of the building confirms Leone's description and data and suggest that he himself was following a detailed and measured plan [figs. 36, 37, 38].[16]

Leone defines the castle *arx*, significantly taking up the distinction made by Leon Battista Alberti in his *De re aedificatoria* between a tyrant's fortress, which had an explicitly military character, and the *regia*, the magnificent residence of

---

11 Leone 1514, bk. II, ch. 1, f. *recto*: "Quonam autem modo nunc se habeat, hoc est anno MDXII post Iesum Dominum, consequens est enarremus".

12 *De elegantia vitaeque cultu atque felicitate Nolanorum.* Leone 1514, bk. I, ch. 15, ff. xxi–xxi *verso*.

13 Leone 1514, bk. II, chs. 1–2, ff. xxii *recto*–xxiiii *recto*. See Lenzo in this volume. Chapters 3 to 6 are dedicated to the modern appearance of the surrounding *ager*.

14 *De arce nolana.* Leone 1514, bk. II, ch. 8, ff. xxix *recto*–xxix *verso*.

15 Leone 1514, bk. II, ch. 8, ff. xxix *recto*–xxix *verso*.

16 Carillo 1996; Caianiello 2003.

the prince, which in the case of Nola was the baronial palace belonging to the Orsini [fig. 28].[17]

The description of the Orsini palace had already been partly introduced in Book 1, in the course of Leone's attempt to reconstruct in words what he took to be the "marble amphitheatre" [figs. 11, 12].[18] The ancient monument had been excavated during the 1460s in front of his eyes, while he was still a child, by the count Orso Orsini, who was redeploying its limestone blocks to build the façade and other parts of the baronial residence.[19] Later on, the unearthing of the monument gave Leone the opportunity to inspect and measure its remains and to attempt a reconstruction of its original dimensions and appearance. As an eyewitness both of the excavation of the amphitheatre and of the construction of the Orsini palace, Leone gives a sense of how the two building sites progressed in parallel and how the project of the residence was altered as the excavations advanced. After presenting the image of a quadrangular and symmetrical palace with sides measuring twenty-five "passi" or steps, Leone emphasizes how initially Orso had begun to build his residence in tufa stone, using the ancient blocks which had come to light to reinforce only the corners of the new palace, "not being aware of the amount of marble which was hidden in the foundations of the eastern amphitheatre".[20] The large number of "marble blocks", which gradually emerged from the site, induced Orso to use them for the entire main façade, thus covering the wall which had already been built in tufa stone, while further blocks were used for pilasters, bases and cornices, leaving many other stones unemployed.[21] Leone used this skilful account of how the façade of the new residence gradually grew as the ancient blocks of the amphitheatre were dug out of the earth to explain why the northern elevation had been executed "ex marmore quadrato", that is, in square marble, while the foundations and the other sides of the palace were

17   *De regia.* Leone 1514, bk. II, ch. 9, ff. xxix *verso*–xxx *recto.* Alberti (ed. Orlandi) 1966, 346–347.

18   Leone 1514, bk. I, ch. 8, ff. xxii *r*–xxii *verso.* For the description of the marble amphitheatre see de Divitiis, Lenzo in this volume, 49–52. On the Orsini palace in Nola see Manzi 1970; Clarke 1996; Stefanile 1996; Cirillo 2002; de Divitiis 2013, 210–217; de Divitiis 2016.

19   Leone 1514, bk. I, ch. 8, f. xii *recto*: "Marmoreum vero, quod nobis pueris magis corruptum atque dirutum erat quam latericium, etiam paulo post fere funditus extirpatum est ab Urso Ursino Nolae tum regulo, Nicolae avi tui patruele". See de Divitiis, Lenzo in this volume 50–51.

20   Leone 1514, bk. II, ch. 9, f. xxix *verso*: "Mens enim erat Urso fabricam omnem ex tophis conficere, ignoranti multitudinem marmorum quae in fundamentis amphitheatri latebant orientalis".

21   Leone 1514, bk. II, ch. 9, f. xxx *recto.*

in tufa stone, providing a further evidence of his direct personal experience of the building work in progress. Leone briefly mentions the marble portal and the marble bust portraying Orso Orsini in a niche at the centre of the facade, but his main interest seems to be in the transport of the blocks from the amphitheatre to the new baronial residence, which served to emphasize how the magnificence of ancient Nola had been transferred to the modern city.[22]

The accounts of the castle and of the Orsini palace both show how Leone grounded his descriptions on a familiarity with the sites and on the existence of measured drawings. This method becomes even more evident in the case of the Seggio, the main public building where the members of the noble élite used to meet, located in Nola's main square [fig. 33].[23] Here Leone outlines the small building as though he were reading a drawing on paper, describing the rectangular plan whose sides measure respectively twenty and five steps; he also adds that the southern and western sides presented two adjoining open arches, both of which rested on a single corner column, and that there were seats running the entire way round the interior space. Even in its brevity, this description affords us an insight into how Leone's descriptions emphasize spatial relations, geometries and measurements.

Such an approach is particularly in evidence in Leone's account of one of the most complex and stratified buildings in Nola, the cathedral dedicated to the Assumption and to the two saints Felix and Paulinus.[24]

3   The Cathedral

The cathedral was erected in the fourteenth century on the site of the former basilica of the episcopal complex, and it was completed in the mid-fifteenth century when—as Leone himself recounts—his uncle was the director of the building work [fig. 33].[25] The description of the cathedral in the *De Nola* is a

---

22   Leone 1514, bk. II, ch. 9, f. xxx *recto*. Leone clearly describes how the construction had begun from the west wall, progressed towards the north side, in other words the main façade made of limestone blocks from the amphitheatre, and then continued towards the east. These parts had been completed up to the roof, while the southern side had only reached the first floor. Leone further notes that remains of a pre-existing building were still visible.

23   *De porticu Nolana atque viis urbis.* Leone 1514, bk. II, ch. 10, f. xxx *recto*. On the Seggio in Nola see Lenzo 2014, 178–179, 200.

24   *De Basilica Urbis principe et Episcopio.* Leone 1514, bk. II, ch. 9, ff. xxxi *recto*–xxxiii *verso*.

25   The cathedral was erected between 1370–1395, most probably on the site of a previous church dating back to the end of the twelfth century. According to recent studies, the

unique source for the reconstruction of the appearance of the building at the turn of the fifteenth and sixteenth centuries before it was affected by further destruction and restoration.

Leone introduces his account by taking up the terminological and cultural debate on the terms *templum* and *basilica*, exactly as Alberti had done in discussing the type of the ancient basilica in the *De re aedificatoria*.[26] He then takes into consideration the orientation of the cathedral and its position within the city, as well as its relation to the other buildings which formed the episcopal complex, and in particular to the bell tower and the early Christian Basilica of Santi Apostoli, both connected to the main building even though situated at a lower ground level.[27]

Throughout the description Leone maintains a basic anthropomorphic metaphor where the apse is described as the head and the entrance portals as the feet, an image which easily allows him to pass from the account of the single parts to that of the whole building. Leone was most certainly aware of how the anthropomorphic interpretation of architecture was one of the central themes of Vitruvius's *De architectura*, and also of how such an idea had been developed by Alberti, who in his *De re aedificatoria* affirms that "a building is a body" and often compares the parts of an edifice to the skeleton, limbs and nerves.[28]

---

new Gothic-style building was sponsored by Niccolò Orsini after the archbishop of Nola Francesco III Saccano obtained in 1370 from Pope Gregory XI permission to transfer the episcopal seat from Cimitile to Nola. It is still not clear if Nola had actually been deprived of its episcopal status in favour of Cimitile meaning that it already had a cathedral and that therefore the fourteenth-century works were a restoration of the twelfth-century building (Ebanista 2007). In 1594 the cathedral was rebuilt after it collapsed in 1583 when a tomb was dug out of a pilaster. Destroyed by fire in 1861, the building was reconstructed in a neo-Renaissance style in 1909 (Remondini 1747, I, 161–168; Carillo 1993; Ebanista 2007, 38, 48).

26   Leone 1514, bk. II, ch. 11, f. xxxi *verso*. Alberti (ed. Orlandi) 1966, bk. VII, ch. 14, 631–634.

27   Leone 1514, bk. II, ch. 11, f. xxxi *verso*. The Basilica of the SS. Apostoli, generally dated to 89 CE, would have functioned as the cathedral until the supposed transfer of the episcopal seat to Cimitile in 409 CE. by Saint Paulinus. Restored in 1190, the basilica was used as the cathedral between 1370 and 1394. The present-day appearance is the result of the restoration carried out by Domenico Antonio Vaccaro between 1735 and 1741, but the original form of the church with three naves separated by marble columns is still recognizable. Angelillo 1909; La Rocca 1971; Angelillo, La Rocca 1971; Avella I, 74–90; Lenzo 2006, 270–329; Russo 1973. On the bell-tower, built between the twelfth and thirteenth centuries see Rosi 1949.

28   Vitruvius (ed. Gros) 1997. Alberti (ed. Orlandi) 1966, I, *Prologue*; bk. III, ch. 14, 14–15; 246–247. For a thorough discussion of the subject, including the connection between medicine and architecture, see Collareta 2008, 167–168. In this context Collareta draws our

At the same time the parallel between the church and the human body expressed in the late thirteenth century by Guillaume Durandus in his liturgical treatise, the *Rationale divinorum officiorum*, was also familiar in humanist discourse.[29] In addition to the anatomical parallel, it is interesting to note how, in order to explain the form of the cathedral, Leone recalls the comparison expressed by Alberti in the *De re aedificatoria* between the plan of the early Christian basilica, with its transept, and the form of the letter "T".[30] He then further enriches the Albertian analogy by stating that the body of the basilica with the apse forms a monk's "tunic with the hood and the sleeves stretched out".[31] This striking comparison, which is more appropriate for the complex structure of the modern cathedral and enhances the visual force of Leone's description, draws on the long-standing symbolic association of the monastic habit with the shape of the cross, often evoked in medieval sources, such as the Franciscan habit described by Thomas of Celano's *Vita Prima* (1228–1230).[32] A very effective pictorial transposition of how Saint Francis drew the shape of his habit as if it were a cross, closely recalling Leone's account of the Nolan cathedral, can be seen in the scene of the *Choice of the habit* in the *Bardi dossal* (1245–1250) in Santa Croce in Florence [fig. 34].[33]

After evoking the visual image of the entire building, Leone analyses the cathedral in detail, describing a sequence of arches on pilasters which divided the building into seven parts: the three naves, the transept, the central apse

---

attention to a passage of Giannozzo Manetti's *De dignitate et excellentia hominis* where, in order to describe the human body, the humanist uses architectural terminology (Manetti ed. Leonard 1975, 25).

29   Durandus (ed. Davril, Thibodeau) 1995, I, i, 4, p. 17. Collareta (2008) also points out the anthropomorphic description of the cathedral of Santa Maria del Fiore in Florence contained in Alberti's *Profugiorum ad aerumna libri tres*. He also stresses the importance of Antonio Manetti's passage in his *Life of Filippo Brunelleschi*, according to which the architect recognized in the ancient way of building a certain order of limbs and bones ("vide el modo del murare degli antichi e le loro simmetrie, e parvegli conoscere un certo ordine di membri e d'ossa"). See Manetti (ed. De Rubertis, Tanturli) 1976, 64–65.

30   Leone 1514, bk. II, ch. 11, f. xxxi *verso*. "Quamobrem hae duae aedes figuram T literae praesentant, cui si cellam superaddideris, tunicam cum cucullo ac extentis manicis finxeris". Alberti (ed. Orlandi) 1966, bk. 7, ch. 14, 633–634.

31   Leone 1514, bk. II, ch. 11, f. xxxi *verso*.

32   Celano (ed. Da Campagnola) 1995.

33   On the Franciscan habit and the *Bardi dossal* see Warr 2010, 100–103. On the symbolism associated with the monastic habit see Constable 1996, 191–192. On the *Bardi dossal* see Frugoni 1988, 9; Boskovits 1993, 112–116: 472–507; Cook 1999, 98–102; Brooke 2006, 176–192; Tartuferi 2007, 50–55.

and the two minor lateral apses.[34] Adopting Alberti's architectural perspective, he draws his reader's attention to the spatial and volumetric connections and to the relations of the seven parts to each other, stating for example that the three arches which connect the transept to the nave created a wide passageway and made the areas of these two different parts of the buildings identical.[35] The description continues with the façade of the cathedral, the "feet" of the body, consisting of a central tympanum and two lateral semi-tympanums and marked by windows and three doors, the central one of which was higher than the others.[36]

Following the same logical approach Leone then describes the spatial implications of the choir at the centre of the major nave with the rest of the building and the relations between the basilica and the crypt below the choir containing the relics of the third-century martyr Saint Felix. Similarly Leone indicates the passage from the southern side of the transept to the adjoining Basilica dei ss. Apostoli [fig. 33].[37] The description of spaces is integrated with a complete set of measurements of the entire cathedral and of its single parts, accompanied by remarks on the materials, as when Leone notes that the roof was made of tiles, the pavement was in marble, and the seats of the choir were made of walnut wood.[38]

Leone does not limit himself to describe the modern cathedral: he also tries to identify its different historical layers and its relation to the other buildings which form part of the episcopal complex and show clear signs of being earlier constructions. In order to reconstruct the original appearance of the site Leone explicitly draws on *Letter 32* written by Paulinus of Nola to Severus Sulpicius in the early fifth century, inserting lengthy quotations from it which, in the context of the *De Nola*, read like what is to all effects a poetic *ekphrasis*.[39] However, the notable description which Paulinus gives did not refer to Nola, but to the four basilicas in Cimitile, the nearby episcopal site built on the tomb of the

---

34   Leone 1514, bk. II, ch. 11, f. xxxi *verso*.

35   Leone 1514, bk. II, ch. 11, f. xxxii *recto*.

36   Leone 1514, bk. II, ch. 11, f. xxxii *recto*.

37   Leone 1514, bk. II, ch. 11, ff. xxxii *recto*–xxxii *verso*. In bk. II, ch. 16 Leone recounts that the choir had been built during the completion works at the time of Orso Orsini, therefore between 1462 and 1479, by the archbishop Giovanni Antonio Tarantino (Leone 1514, bk. II, ch. 16, f. xxx *recto*). For rood screens and choirs see Cooper 2000. On choirs in southern Italy see Scirocco 2009.

38   Leone 1514, bk. II, ch. 11, f. xxxii *recto*.

39   For Paulinus' letters see Goldschmidt 1940; Quasten 1996, Trout 1999.

martyr Saint Felix, where Paulinus had erected a new basilica [fig. 35].[40] It is difficult to ascertain the reasons behind Leone's erroneous use of the description of Cimitile in his discussion of the original appearance of the cathedral in Nola. That Leone knew Cimitile well is clear from the long account dedicated to the site in Book 1 and we cannot exclude the possibility that he used Paulinus' text on purpose, in order to underline the ancient origins of Nola as an episcopal seat. But it is also possible that the error arose because Paulinus himself refers to the basilicas in Cimitile as being in Nola, thus reflecting the controversial dual role of the two episcopal sites and also the existence of two different San Felices who were venerated respectively in the two centres.[41] Yet a further possibility is that Leone strategically used the description of Cimitile to create in his reader's mind the image of the appearance that the first cathedral of Nola might have had in antiquity, lavishly decorated, with coffered ceilings, columns and inscriptions. Be that as it may, Leone clearly wanted to match the text from the letter of Paulinus with the appearance of the *insula episcopalis* in Nola, arguing that the origins of this site went back to the ancient temple of Jupiter, the remains of which could perhaps be recognized in the crypt of the modern cathedral.[42] Even though Paulinus' text led him to an erroneous interpretation of the history of the city's cathedral, his use of the ancient letter can be seen as evidence of how Leone recurred to the primary sources at his disposal in order to interpret the evidence he had under his eyes.[43] At the same time, though, it is interesting to note that Paulinus' account was not used as a model for Leone's own method of architectural description, which seems to be without precedents.

Leone's description of the cathedral stands out as exceptional for his time, when the dominant model for accounts of religious buildings still focused on saintly relics, miracles and liturgical furnishings, rather than on space. It is also far removed from the lists of churches and chapels in Rome compiled by Biondo Flavio in his *Roma instaurata* (1446–1448) or later by Francesco Albertini in his *Opusculum de mirabilibus novae at veteris Urbis Romae* (1510).[44]

---

40   Leone's extensive description of Cimitile is in bk. 1, ch. 14 (Leone 1514, bk. I, ch. 14, ff. xix *recto*–xix *verso*. For Cimitile see Ebanista 2007 with previous bibliography.

41   Paulinus mentions the *Nolana basilica* (*ep.* 32.17). Cfr. Leone 1514, bk. II, ch. 12, f. xxxiii *verso*–xxxiiii *recto*. On the two episcopal sites and on the two San Felice, San Felice prestibero and San Felice martire see Ebanista 2007.

42   Leone 1514, Bk I, ch. 14, ff. xviiii *verso*–xx *recto*; Bk II, ch 12, ff. xxxiii *verso*–xxxvi *recto*. On the Basilica of the ss. Apostoli see footnote 29.

43   For the judgments expressed by eighteenth-century writers concerning the unreliability of Leone's work on Nola see Ruggiero 1997, 82–84.

44   Biondo 1471; Biondo (ed. Raffarin-Dupuis) 2005–2012; Albertini 1510.

Two architectural descriptions which were surely important for Leone were those of the basilica of San Marco given by Marcantonio Sabellico first in the *Decades rerum Venetarum* (1486–87) and then in the *De situ Urbis Venetae* (1491), even though the humanist mainly focuses on the rich decoration, the materials used, the techniques adopted in the paving, the columns, the marble revetments of the walls and the golden vaults, rather than on space.[45] A partial precedent can be traced back to the description of the cathedral of Teramo written by the humanist Giovanni Antonio Campano around 1460, which shares the same spatial approach we find in the description of Nola's cathedral: it is not known if Leone was aware of Campano's account, but it gives us an insight into the architectural knowledge which could have been shared by two humanists with a southern upbringing.[46]

Yet compared with these precedents Leone goes further. His description of the cathedral focuses on the spatial aspects of the building, passing from the overall form to the single parts and finally to the relation between each of the parts and then between the parts and the entire building. He does look at materials, though he describes them as an architect would, rather than from a devotional point of view. Without lingering on the sculptural details and on the liturgical furnishings, Leone concentrates on measurements and proportions, in a type of description which suggests that he was relying on drawings or at least on annotated sketches of plans and possibly sections of the edifice, as well as a direct knowledge of the building site. As in the case of the excavations of the amphitheatre and the construction of the Orsini palace, it is possible that his local contacts, especially his uncle who had been director of works, and other figures involved in the city's architectural projects, acted as a privileged source of information for him. Leone was certainly not constrained in choosing to concentrate on space and geometry while making no reference to the many works of art adorning the building.[47] This choice of approach seems even more significant when we consider his notable skills in describing even the most minute works of art, as in the *ekphrasis* in the *De nobilitate* on the precious inkwell made by Caradosso Foppa (1452–1526 or 1527) for Cardinal Giovanni d'Aragona [fig. 58].[48] In the case of the cathedral, any reference to the decoration would have been a distraction from what appears to be the author's

45 For Sabellico's architectural descriptions see Modesti 2009, 22–25.

46 For Campano's letter see Campano (ed. Menckenius) 1707, 25–35.

47 For altars, monuments and liturgical furnishings, among which the notable Easter candle commissioned by the Orsini and a jeweled cross see Guadagni (ed. Toscano) 1991; Toscano G. 1996; Campone 2000; Campone 2009.

48 See in this volume Lenzo, 68, 79; Loffredo, 111–115; and Appendix 5, 170–171.

primary interest, that is the verbal recreation of the actual physical appearance of the building, in a re-evocation of the type of structured and terminologically precise discourse adopted by Alberti for the description of architectural projects.[49]

4   The Nolan *domus*

Leone's architectural expertise and spatial approach culminates in the chapter dedicated to the Nolan *domus*.[50] In order to give an idea of the residential mansions in Nola, rather than describing a single building or making a survey of the most remarkable houses of the city, Leone takes on the difficult and sophisticated task of defining a general and abstract model of the Nolan house. It is an outstandingly original operation in which Leone merges his direct knowledge of many different examples of Nolan houses with his theoretical understanding of the numerous literary models which dealt with the theme of the house and with exceptional research into the appropriate terminology.

The chapter comes towards the end of Book 2, after the descriptions of the single most prominent edifices in the city, and it precedes the section on the worthy men who had commissioned buildings in Nola from antiquity up to Leone's own times.

The title of the chapter, "The form of the houses in the city of our days and the disposition of their parts" ("Quae sint figura aedium praesentis urbis et qualiter earum partes se habeant") tells us immediately that Leone will analyse the house in parts. After having indicated its general form as quadrangular, he emphasizes the almost constant presence of the courtyard (*cavaedium*), which could be bounded on one or more sides. He then makes a list of all the parts and provides a detailed and systematic description of every single space. In the description of the house, as in the previous examples, Leone proceeds from the general to the particular and, almost as if he were leading a visitor by the hand, he advances from the lower to the upper floors.

The description begins with the entrance, which he calls *atrium*. After having lingered over the nearby stables ("equile quam stallam vocant"), Leone describes the cellar ("cella vinaria [...] quod cellarium vocant"), built at a distance from the entrance, writing at length on the shape of the wine barrels and the etymology of their name. He then singles out the presence of the puteal in the courtyard, which he asserts had been used since ancient times to supply

---

49   Alberti (ed. Orlandi) 1966, vol. II, bk. VI, ch. 1, 440–444. See Burns 1998, 120–123.
50   Leone 1514, bk II, ch. 15, ff. xxxviii *recto*–xxxix *verso* (see Appendix 3, 166–169).

water to individual houses by collecting it from the numerous natural springs underground. The account of the ground floor is completed by the indication of the possible presence of an oven (*furnus*) and of a wash-house ("lavacrum quod cantarum pannorum vocant") in the courtyard, and of a walled garden at the back of the house [fig. 49]. A staircase located either in the atrium or in the courtyard leads to the first floor. Here there was the dining room ("conclave autem vel Graece dicas triclinium, salam Nolani vocant"), with two windows overlooking the road, which eventually adjoined a *loggia* or portico ("porticus superior quam logiettam, hoc est scaenulam vocant"), open on one or two sides with arches resting on pilasters or small columns. Next to the hall there were 32-foot square bedrooms ("cubicula quas cameras appellant"), and a *studiolo* ("bibliotheca quod studium vocant") of the same dimensions, or somewhat smaller, where the books were arranged on shelves fixed to the walls (*plutei*). Next to the dining rooms there was also the kitchen (*coquina*) and the pantry ("penuaria cella quam spensam vocant") [fig. 50]. Leone moves then to the upper warehouse (*megakenon* or *navis*), an open room with small windows, covered by a roof consisting of a single or double slope specifically designed for housing the *avellae*, the local type of hazelnuts [fig. 51]. Leone ends his account by enumerating all the elements that were instead missing in the typical Nolan *domus*, such as towers, a third floor, a terrace and shops on the ground floor. This is a clear indication that he was comparing the Nolan *domus* with other models of palaces he would have seen in other cities such as Naples and Venice, where he had lived, as well as Padua, Rome, Salerno, Siracusa and Messina which we know he had visited.[51] It also tells us that Leone was writing for readers who would have expected to find these features in his description of the house.

Rather than listing items of furniture and other furnishings, Leone focuses on the geometric form and on the proportional relationships of the rooms, indicating for example that the *sala* is rectangular, and twice as long as it was wide, or that the bedrooms are always square. The sole significant reference to the internal furnishing is to the *plutei* in the *studiolo* on which the books were shelved, a detail which has a social and cultural relevance suggesting that the *studia* and books are given a place of pride, and that most of the Nolan families were cultivated. In the same fashion Leone's emphasis on the wooden

---

51 From the *De nobilitate* it is clear that Leone knew Salerno and Rome; see Leone 1525, ff. h iii *verso*. That Leone compared the palaces of his city with those seen elsewhere is confirmed by the fact that in a later work, the *Novum Opum Quaestionum*, the humanist states that the palaces of Messina were very similar to those in Nola; see Leone 1523, *problema* 288. Siracusa is mentioned in the *De Nola*; see Leone 1514, bk. III, ch.2, f. xxxxii *verso*.

wine barrels in the cellar off the courtyard, dignified by an improbable Greek etymology, and on the sacks of nuts kept in the upper storey, can be seen as an indirect way of stressing the wealth of the local citizens who, as described in Book 3, derived their income from their productive and fertile estates and through commerce of these products.[52]

Leone claims Nola could boast seven hundred noble houses, but he makes no specific reference to any of them: apart from the exceptional case of the Orsini palace, details relating to single houses in the entire text of the *De Nola* are purely incidental.[53] It is clear, however, that the description of the *domus* is the result of a study of many different houses which were known to him in Nola and in the surrounding areas, and the features he describes can still be traced in those fifteenth and early sixteenth century palaces which still survive in the city and in nearby centres, such as the Ceva Grimaldi, Mastrilli and Albertini palaces in Nola, the Petrucci palace in Carinola and the Azzolini palace in Aversa [figs. 32; 41–47].[54] As in the cases of the previous building, in order to describe the model of the Nolan house Leone would have most

---

52   See Vitale in this volume, 129–130.

53   For the estimate of noble houses in Nola: Leone 1514, bk II, ch. 15, f. xxxix *verso*. When reconstructing the appearance of the ancient city Leone mentions how the houses of the Cesarini, Mazzeo, Frezza and Tango families had been built on an ancient solid wall of great width (Leone 1514, bk I, ch. 8, f. xiii *verso*). On another occasion, while discussing the presence of Jews in Nola, Leone describes how the surgeon Guglielmo had dared to widen and refashion the large house on the Via Vicanziana, one of the most important roads of the centre along which the most privileged palaces were located (Leone 1514, bk III, ch. 4, f. xxxxix *verso*).

54   In his description of two different types of possible staircases, one in the atrium and one in the courtyard, Leone seems to be recording the evolution of this feature over time in Nolan architecture, by comparing for example the early fifteenth-century staircase in the atrium of the Ceva Grimaldi palace (15th century), with the new type introduced in the new Orsini palace in the 1470s, and later imitated in the palaces of the Albertini (c. 1470) and Mastrilli (15th–16th centuries) families. The sequence of the atrium and the quadrangular courtyard can still be observed in the fifteenth-century palaces in Via Santa Chiara. Avella 1996–1999, III, 512–513. When Leone states that the ceiling of the atrium usually consisted of wooden beams and was only rarely vaulted, he seems to be comparing the exceptional example of the barrel vault in the Orsini palace with the entrances which were more common in Nola, and which can still be seen in the palaces which line Via Santa Chiara. The storeroom for nuts, with its characteristic sequence of small windows, can still be seen in Nola in a palace in Vico Santa Marta and further examples can be found in the palaces of the nearby city of Aversa. The loggia adjoining the *sala* on the first floor can still be found in the Petrucci palace in Carinola, as well as in the palace in via Monferrato and the Azzolini palace in Aversa. On the Orsini palace see footnote 18. On the Albertini

probably availed himself of measured drawings of palaces, either made by himself or easily available to him. Leone possibly drew also a diagrammatic plan where he summarized the main common elements detected in the many residences he analyzed: following his words we can actually redraw the plan of a house.

The abstract character of the description is brought out by the fact the Leone considers different and alternative solutions for the arrangement and dimensions of many features: for example, he indicates that there could also be additional dining rooms or different dimensions of the atrium depending on the width of the house and the magnanimity of the owner.[55] Just as we have seen in his account of Nola's cathedral, so too with the *domus* Leone describes the spatial relationships between the entire building and its single parts, following a precise hierarchy in the way they are distributed, and highlighting the relationships between spaces and floors, as when he indicated the courtyard as the physical centre of the house and recognizes the dining room as the space to which all the rooms on the first floor relate. He also provides information on the materials used, stating that houses were made of tufa stone rather than brick, that the pavement of the vestibule was composed of small lava flags and that the ceilings between floors were constructed of walnut beams, thus confirming again that his architectural knowledge was not confined to theoretical matters, but also included a familiarity with practical building.

While Leone's architectural knowledge and ability to construct a clear argument relating to buildings emerge in many parts of the *De Nola*, it should be emphasized that the decision to create an abstract description of the model of the Nolan house is far from being an obvious choice. Compared to the literary descriptions of buildings developed by other humanists of his own and the previous generation, Leone seems remote both from Biondo's archaeological approach to the concept of the house, and from the paratactic descriptions of Marin Sanudo and Paolo Giovio, who described specific residences by listing the different spaces one after the other and focusing on the furniture of the interiors.[56] Compared to the abstract model of a cardinal's house created by Paolo Cortesi in book 2 of the *De Cardinalatu* (1510), Leone's description of the

---

palace see Mollo 1996. For the Petrucci palace in Carinola see Robotti 1979; Cundari 2005. For the palaces in Aversa see Amirante 1998, 150–158; Fiengo-Guerriero 2002, 360–362.

55   Leone 1514, bk II, ch. 15, gg. xxxviii *recto*–xxxix *verso*.

56   On Biondo's discussion on the ancient house and its furnishings in Book IX of the *Roma triumphans* see Burns 1998, 121; Clarke 2003, 97–98. On Sanudo's description of the Castel Nuovo and Castel Capuano in Naples see de Divitiis 2011. On Giovio's descriptions of Poggioreale and Duchesca see Maffei 1996; Maffei 1999.

Nolan *domus* omits all references to the palaces he actually saw for the sake of a total abstraction.[57]

The title of the chapter and the way in which Leone develops the subject in single parts seems to reveal an awareness of Francesco Mario Grapaldi's *De partibus aedium* (1494), a best-seller of Leone's time, which explicitly aimed at describing not the house as a whole but just its single parts; it was a kind of hand book and lexicon which, intending to be useful for the public, analysed the terms relating to each space and the evolution of each feature over time.[58] The same kind of method was adopted also by Raffaele Maffei, who in his encyclopaedic *Commentariorum urbanorum libri XXXVIII* (1506), devotes book 27 to the theme of the house, entitling it significantly "De domo et eius partibus".[59] But the way Leone identifies common features, organizes his argument and selects a specific terminology, reflects above all his study of the sixth book of Vitruvius's *De architectura* and of the fifth and ninth book of Alberti's *De re aedificatoria*, together with an up-to-date knowledge of the antiquarian and architectural debate on the theme of the house which had been continuing for over fifty years.[60]

It is interesting to note that Leone had already confronted the theme of the house, in generic terms, some years earlier in the *De nobilitate rerum dialogus*, where he briefly celebrates domestic residences as the most commodious work of architecture.[61] Compared to this precedent, the chapter on the *domus* in the *De Nola* is far more complex, revealing the amount of research and reflection on the theme of the house which he must have undertaken in the years between writing the two descriptions.[62] In this frame of time Leone most probably had the opportunity to study Vitruvius more effectively thanks to the appearance of the new edition of the treatise edited by Fra Giocondo, published in Venice in 1511, three years before *De Nola* was issued. Fra Giocondo had been able to provide a philologically acceptable interpretation on some critical points and lexical *cruces* which had perplexed humanists and architects over previous decades, such as those relating to *atrium* and *cavaedium*, which surely made Leone's own studies slightly easier than his predecessors'.[63]

---

57   Cortesi 1510. See Weil-Garris Brandt, D'Amico 1980; Giannini 2003; Chambers 2009.

58   Grapaldi 1494. See Clarke 2003, 98–100; 106–107.

59   Maffei 1506; Clarke 2003, 100.

60   On the study of Vitruvius and Alberti among humanists see Clarke 2003, 100.

61   Leone 1525, f. hiiii *recto*.

62   Leone 1525, ff. hiiii *recto*.

63   Vitruvius (ed. Giocondo), 1511. On the *domus* described by Vitruvius see Clarke 2003, 92–93; Gros 2015. On the difficult interpretation of ancient terms relating the *domus* see Clarke 2003, 85–125.

Furthermore the idea of rebuilding *the* Nolan home was probably inspired by Fra Giocondo's work model of the plan of an ancient aristocratic *domus* which gave for the first time a systematic overview of the main elements which made up ancient houses, such as the *atrium, cavaedium* and *cubicula,* as well as the *bibliotheca* [figs. 49, 50].

Even more than Vitruvius, the main source for the chapter on the Nolan *domus,* as well as for many other aspects of the *De Nola,* seems to have been Alberti, of whose general concepts and also some detailed readings we find echoes in Leone's work.[64] The idea itself of having a specific chapter on the house within a work about a city seems to put into effect the Albertian analogy according to which the house is like a small city and the city is like a large house. Such an analogy makes the treatment of the *domus* in the description of a city, such as the *De Nola,* indispensable.[65] The title of Leone's chapter "The form of the houses in the city of our own day and the arrangement of their parts," reveals a way of treating the argument by proceeding from the general to the particular, a homage to the concept of *concinnitas* that underpinned Alberti's entire work, as was the idea of discussing the relationship between the whole and the single parts and the relation between the parts among themselves.[66] Looking in more detail, Leone draws above all on the features of the noble villa described in Chapter 17 of Book v of the *De re aedificatoria.* Even if he does not use Alberti's term *sinus* for the courtyard, Leone considers the *cavaedium* to be the main part of the house and the one that determines its general form.[67] A proof of Leone's dependence on Alberti's text is given when, in recalling the etymology of the word *sala,* which was the Nolan term to indicate the dining room/living room on the first floor, the author reproposes the exact explanation of the term given in chapter 2 of Book v of the *De re aedificatoria,* writing that the word *sala* is derived from the Latin verb *saltare,* since the room was used for dancing during banquets and weddings.[68] Leone thus borrows from

---

64 As Carlo Vecce points out, Leone's study and use of Alberti's *De re aedificatoria* was paralleled by Jacopo Sannazaro who in the same years compiled an index of terms and topics from the architectural treatise. See Vecce 2000; Vecce 1996 (1997). On Leone and Alberti see also De Filippis 1991.

65 Alberti (ed. Orlandi) 1966, vol. I, bk. I, ch. 9, 64–65. On Alberti's influence on Leone see De Filippis 1991; Borsi 2006, 295–305.

66 On Alberti's *concinnitas* see Alberti (ed. Orlandi) 1966, vol. II, bk. IX, ch. 5, 814–815; Burns 1998, 124.

67 On the use of the Albertian *sinus* in Sannazaro, see Biermann 1990.

68 Alberti (ed. Orlandi) 1966, vol. I, bk v, ch. 17, 414–432. On Alberti's vocabulary see Biffi 2007; Lücke 1975–1979, 173–174; 236; 1202. Leone 1514, bk. II, ch. 15, ff. xxxviii *recto*–xxxviii *verso*. On the *atrium* Leone's phrase closely resembles Alberti's (bk. v, ch. 2,): "Intra aedes

Alberti the same false Latin etymology for the Italianism *sala*, a word which instead had Lombard origins.

Apart from the reference to Alberti, the latter examples reflect Leone's extensive research into the lexicon of architecture and his knowledge of the nuances of some controversial terms, such as *atrium* for the entrance, *cavaedium* for the courtyard and *cubicula* for the bedrooms.[69] That Leone referred to a specific architectural terminology is admitted by the author himself in another part of his work where he describes an ancient underground porch ("subterranea porticus, antiquum opus") located between two religious foundations near Nola and explicitly specifies that some architects called such structures *cryptoporticus* ("quam cryptoporticus architecti quidam vocant").[70]

Leone's terminological research leads him to investigate the correspondences between different languages, so that almost every Latin term is followed or sometimes introduced by an equivalent in Greek or in Nolan, or both. In the simpler cases Leone creates a correspondence between humanistic Latin and vernacular terms in a Latin form, stating for example that *cubicula* are called *cameras*. On other occasions he seeks out the Latin name for features which did not have an ancient precedent or which had only local connotations, as when he mentions that the little fragments of lava known as *lapilli* might be more properly termed *parvae pumices* and that the paving of the vestibule made of *lapilli* was normally referred to using the Greek word *astracum*.[71] In some cases, Leone uses Greek to make up etymologies for terms that had been in use only since medieval times, as when in discussing the form of the storeroom he writes that the Nolan word *magacenum* comes from the Greek *megakenon*, later interpreted in Latin as *magnum vacuum*.[72] Something similar

---

ambulationem aream atrium salam, quam a saltando dictam puto, quod in ea nuptiarum et convivarum alacritas celebretur, non universorum sed potius incolarum sunt". Alberti 1966, vol. 1, bk v, ch 2, 338–339 (f. 70 *recto*) Leone's profound interest in the vocabulary of Alberti's treatise emerges also in his other works, such as the *De nobilitate*, where he uses *concinnitas* to define the beauty of the Egyptian pyramids (Leone 1525, f. h ii *r*), or in the *Novum opus Quaestionum* where he discusses the problem of the *entasis* of a column calling it "the belly of the column." (Leone 1523, *Problema* 19).

69   For a discussion the terms relating to the *domus* see Clarke 2003, 116–125.

70   Leone 1514, bk. II, ch. 15, ff. xxxviii *recto*–xxxviii *verso*. On the discussion of this passage and its possible references to Alberti see de Divitiis and Lenzo in this volume, 55.

71   On the term *astraco* or *astreco* and its persistence in Neapolitan architectural vocabulary see De Ritis 1845, 191.

72   On the use of *megacenon* in the Middle Ages see Du Cange, Favre 1883–1887, vol. v., col. 326c.

happens with the *porticus superior*, where Leone tries to trace the origin of the vernacular term *loggia* back to the Greek word *logos*.[73] Apart from these bilingual matches, Leone often recurs to trilingual forms, as when in discussing the vernacular term *sala* he indicates its correspondence to the Latin *conclave* and also to the Greek *triclinium*. While such a display of erudition is appropriate in the context of the historiographical aims of the *De Nola*, it can be said that Leone's linguistic research into architectural terminology parallels the pioneering efforts carried out in those same years by architects and humanists such as Francesco di Giorgio, Fabio Calvo and Raphael, who in attempting to translate Vitruvius were often faced with uncertainties about the meaning of certain words relating to the ancient *domus* and how to translate them in terms of contemporary houses.[74]

The decision to devote a chapter of his work to the Nolan house and the accurate selection of the vocabulary he employs suggest that, by describing the *domus*, more than attempting to provide a complete picture of the physical appearance of the city, Leone wishes to give a sense of the existence of an elevated way of social life and to discuss the nobility of the Nolans. By reading between the lines, after all, the real subject of the chapter is not the houses themselves but their owners, whom the reader can imagine while they organize dancing feasts in the *sala*, cultivate hazelnuts, preserve wine barrels in the courtyard, enjoy conversations in front of the view and in the fresh air of their loggia, and devote themselves to humanistic *otia* in the library full of books displayed on the shelves.[75] Having travelled across Italy, Leone was aware that in other Italian cities a nobleman's palace was seen both as a manifestation of the patron's magnificence and as an ornament of the city, and he actually presents the Nolan house as the highest manifestation of the nobility of character of its inhabitants. In this context the decision to develop an abstract model of the local dwelling, rather than describing selected examples of palaces, emphasized how nobility was widely shared among the inhabitants of Nola, and how the city was populated not with a mere handful of magnficent palaces but by many examples of them.

It was no accident that Leone experimented with a similar abstract description of the Nolan funerary chapels, parallel the one he gave of the *domus*, providing in Book 3 an account of the recurring features that formed the *domus*

---

73   On the use of the term loggia see Biffi 2002.

74   On the use of the vernacular in the Renaissance architectural lexicon see Nencioni 1995; Biffi 2010.

75   See Vitale in this volume, 129–130.

aeterna nolana.[76] If the case of the Nolan *domus* is an implicit way of describing the customs and ways of life of the inhabitants of the city, the chapels are the occasion to describe their funerary practices, from the moment of death and the subsequent funerary rituals, similar to those performed for members of the nobility and the royal family in Naples, up to the burial customs and the modes of mourning.[77] The house and the chapel were therefore ways of demonstrating how the Nolans of Leone's time were accustomed to live, and also to die, and of emphasisng the nobility of contemporary Nolan society, in order to conserve its image for posterity.

5   Architecture and Nobility

The notable attention dedicated to buildings in the *De Nola* reflects Leone's idea of architecture as one of the highest expressions of nobility. The choice to concentrate on space and geometry and to elaborate abstract models seems to respond to the need to communicate, by using only words, the image of this nobility to those contemporaries of Leone who had never visited the city for themselves and, above all, to posterity.

The idea of architecture and nobility which underlies the architectural descriptions and in general the entire work can be better understood by considering Leone's theoretical elaboration in his *De nobilitate rerum dialogus*, published posthumously, but probably written years before the *De Nola*, while he was still living in Nola.[78] In particular, the importance devoted to architecture in the work on his native city seems to continue and apply the ideas he posited in his treatise on nobility. Leone himself was not a nobleman and the system he conceived did not identify nobility with family or blood, but rather as a product of the substantial qualities of each entity, of their virtue and perfection. This was true for inanimate materials like minerals and metals, but also for plants, for animals, for human beings and, last but not least, for human products like works of art. Taking a position in the contemporary debate on the *paragone* of the sister arts, in the *De nobilitate* Leone established the primacy of architecture over painting and sculpture and provided an original contribution to the discussion, arguing that the superior nobility of architecture relied on intelligence, on materials and on utility ("summum in-

---

76   Leone 1514, bk III, ch. 4, f. xxxxiiv *verso*.

77   On the funerary practices of the Neapolitan elite and Aragonese royal family see Vitale 2002.

78   See de Divitiis, Miletti 2016.

genium, nobilis materia et praestans commoditas operis"). He also added that beauty in architecture is on a grander scale, because of its dimensions. Finally, compared to the other arts, architecture has a more solid theoretical approach, since a process of knowledge and precepts underlay it as a discipline.[79] The argument is supported by a considerable number of examples ranging from ancient monuments, such as the Egyptian pyramids and the Roman temples, to medieval works, such as the ninth-century Lombard aqueduct still visible near Salerno, and to contemporary structures, like the Gran Sala in Castel Nuovo in Naples [fig. 7].[80]

Therefore, architecture was not simply a form of magnificence, but above all the expression of the nobility of mankind. In accordance with this belief, the public and private monuments which had been built from antiquity up to 1512 described in Book 1 and 2 of the *De Nola* are represented as the highest manifestations of the elevated standard of living enjoyed by the inhabitants of Nola over the centuries and of their enduring nobility of character. According to Leone's idea of nobility, the excellence of Nola's ancient remains and modern buildings was a proof not of the nobility by birth of their patrons and owners, but of the way in which those patrons and owners had developed to the highest levels their natural skills, and in particular their distinction in civil virtues, in the liberal arts and also in the mechanical ones. This view is confirmed in Book 3, where Leone selects one hundred and fifteen among the noblest families of Nola, not because of their origins, but for the intellectual or commercial skills of their members. Interestingly enough, the humanist develops a social topography dividing the noble families of Nola according to the position of the palace in the city. In the relation between architecture and nobility the theme of the house becomes of crucial importance.[81]

In the *De nobilitate* Leone praised the *domus* as the most commodious product of architecture, describing it through a paratactic list of its spaces and functions. In the *De Nola* this generic celebration of an architectural typology is developed into the much more refined and skilful description of the Nolan noble house. In both works the *domus* is the constituent cell of the urban

---

79   Leone 1525, chs. 39; 42, ff. h i recto: *Enumeratio plurium perfectarum artium, et rationes quibus ars nobilitatur; De ingenio materia et commoditate Architectonicae, quibus ars perfectissima et omnium nobilissima probatur.* For a full transcription of the chapters relating to architecture see de Divitiis, Miletti 2016. On Leone and the *paragone* among the arts see Collareta 2006.

80   Leone 1525, ch. 42, f. h iii *verso*.

81   See Vitale in this volume 132–137.

universe and represents the point of intersection between the nobility of the families and the nobility of the city.

The *De nobilitate* culminates with the description of an abstract model of the city which Leone considered to be an architect's most noble achievement, since only an architect is capable of designing both the whole and its constituent parts. By systematically addressing problems of climate and geographical location, infrastructures, public and private buildings, he outlines a remarkable and indeed unique project of the ideal city, whose nobility derived both from its physical appearance and from the character of its inhabitants.[82]

The image of the city described in the *De nobilitate*, populated by noble citizens, enclosing within its walls beautiful public buildings and a multitude of noble houses, is exactly the one Leone recreates when describing his native city in the *De Nola*. Through his exceptional capacity to evoke in words the images both of ancient and modern structures, such as the amphitheatres, the Orsini palace, the cathedral, and the Seggio, he manages to communicate to readers the magnificent appearance of his native city. But it is by recounting the many families which inhabited those "beautiful palaces" and "magnificent houses" that Leone gives a real sense of the nobility of Nola.[83]

---

82   Leone 1525, ch. 43, f. h v *recto: Manufacta sunt ignobiliora naturalibus, et eorum nobilissima urbs est.*

83   Leone 1514, bk. 1, ch. 15, f. xxi *verso.*

CHAPTER 5

Ambrogio Leone and the Visual Arts

Fernando Loffredo

This chapter aims to reconstruct the intense and significant relationship between Ambrogio Leone from Nola and the artistic production of his time both in the Kingdom of Naples and in Venice, his adopted city. Leone's *oeuvre* is particularly recognized for its exceptional antiquarian knowledge of which he shows his mastery in *De Nola*.[1] However, it is possible to read between the lines of his texts how consistently he observed and was aware of the contemporary evolution of the visual arts, especially at that magnificent historical moment in which the rise of Humanism reached the highest level of devotion to the *antichi*.

1 Sculpture Appealing to Poetry: *Beatricium*

It is essential to remember that Leone was also a patron, not only of works of art but also, and in a very special way, of a close dialogue between sculpture and poetry. According to Leone himself, he commissioned Tommaso Malvito, a Lombard sculptor active in Naples, to make a marble portrait of his platonic beloved, Beatrice de' Notari, also from Nola. Leone inserts this information in the third book of *De Nola*, which concerns the *élite* families of the city. Writing about the de' Notari family (the last one he speaks about, significantly immediately following his account of the Leone family), he enlarges the topic by creating this sophisticated *ekphrasis*:[2]

> Lastly, the Notaro family resides in this neighbourhood, among whom Michele and Carlo were knights and the two sons of Carlo had military careers. In the same way, Francesco, Roberto and his son Bernardino were worthy senators; moreover, Silvestro was honoured with the title of canon. In this family the beautiful maiden Beatrice also shone forth. No man, not even Paris, would be capable of judging whether she was

---

1 See de Divitiis, Lenzo in this volume, 45–58.
2 On Leone's literary skill and attitude to virtuoso descriptions see Miletti in this volume, 11–41, and on architectural *ekphraseis* see de Divitiis in this volume, 81–102.

more beautiful in her visage or in her mind. Both were unparalleled, beyond the most praiseworthy human beauty. The most excellent sculptor Tommaso Malvico [*sic*, for Malvito] sculpted her in marble. Men believe that the Graces and the Horae were present when it was carved, illuminating the artist's eyes and guiding his hands, since she was shown in the Parian marble with such wonderful skill. Indeed, if the marble statue and the living woman had been placed side by side, it would have seemed that there were two Beatrices, one listening, the other speaking; one full of joy, the other laughing. The grace and splendour of the maiden did not only inspire Malvico to sculpt her, but also impelled all the Italian poets to sing her praises, among whom there was Antonio Tebaldeo from Ferrara, a man as extraordinary for his divine intellect and for his speech as he has been celebrated continually everywhere by the most learned of scholars. Thus this double image of divinity was celebrated in innumerable compositions with its epithet bestowed by the Muses, since all praised her as the divine Beatrice. This young girl, at the age of eighteen, went to the beautiful hills of Cicala and while she climbed among olive— and oak trees, was suddenly—wondrous event!—carried away, and could not afterwards be found by anyone in any place. It is believed that, on account of her excellent and divine ways of life, she flew off with heavenly beings while still alive and withdrew herself to the highest heaven.[3]

Thus, Leone commissioned a marble portrait of Beatrice, which was almost certainly a bust—as we will see—and called on several poets to celebrate in verse the beauty of the statue, which was a sublime reflection of the maiden's beauty. On the other hand, Leone contributes to the mystery surrounding this young girl by relating her—uncanny—disappearance in the country setting of Monte Cicala, a hill near Nola. This event obviously brings to mind Ovid's *Metamorphoses* and could also conceal metaphorical meanings, but it seems that something of the kind actually took place.

By considering the poetic responses, at least those we know of, it is possible to reconstruct the network of cultural circles with which Leone was connected. It is moreover interesting to notice the poetic publicity which the sculptor, Malvito, enjoyed, thanks to this singular *certamen*.

---

3 Leone 1514, bk. III, ch. 3, f. xxxxix *recto*.

In order to analyse in depth this testimony of the now lost sculpture,[4] it will be useful to chronicle its critical fortunes, adding at the same time some considerations. Probably the first art historian who noticed this excursus was Leopoldo Cicognara in his monumental *Storia della scultura italiana*.[5] He mentions "Tommaso Malvico" as an unknown Nolan sculptor who carved the statue of Beatrice, which was very famous at that time, at least according to literary sources. Cicognara reconstructs the story of the portrait, basing his account only on Leone's own description and on the sonnets by Antonio Tebaldeo which he was able to find, transcribing both in a long footnote.[6] Several years later Erasmo Pèrcopo linked the references to "Malvico" provided by Cicognara to the actual and very active sculptor Tommaso Malvito, who came from Como and spent most of his career in Naples.[7] Pèrcopo also published five additional sonnets by Tebaldeo dedicated to the "Beatrice marmorea", since all seven poems were printed in a miscellaneous collection in 1499.[8] Furthermore Tebaldeo mentions the "simulacrum marmoreum divinae Beatricis" in an undated letter addressed to Leone;[9] Tania Basile also noticed that in the same

---

4 As we will see below, Damianaki (1998, 368, n. 14; lastly in 2008, 186–193, n. 6) suggests an identification with the female bust in the Musée du Louvre normally attributed to Francesco Laurana. This identification does not seem convincing to me.

5 Cicognara 1823–1824 [2nd edition], IV, 409–412. The first edition, in three volumes, was published in 1813–1818. Before Cicognara, Muratori (1706, 414) had already noticed, scornfully, three of Tebaldeo's sonnets for Beatrice, but he thought they were praising the figure of a saint and not the statue of a young woman.

6 Cicognara had an authentic scholarly interest in the—for him—totally unknown "Malvico", since he writes that "it would be useful to discover if his works of art correspond to the fame that the writers of his times accorded him" ["Uno scultore di cui sarebbe grato il poter iscuoprire se le opere corrispondano alla celebrità che gli diedero gli scrittori del suo tempo"].

7 Percopo 1893.

8 The seven sonnets are: *Che guardi e pensi? Io son di spirto priva*; *Tu che mirando stupefacto resti*; *Che non pò l'arte? Io scio che sei lavoro*; *Pietra a chi il ciel representar die' in sorte*; *Se da caldo et humore il viver viene*; *Costei che viva in bianco sasso miri*; *Ben fusti acorto acompagnare il verso*; in Thebaldeo 1499. It is the first edition (see Sutermeister 1946, 37, n. 63; and lastly Basile, Marchand 1989, 91), of a subsequently republished collection. The most recent critical edition of the seven sonnets is by Basile 1992 (texts: tomo I, 354–361, nn. 223–229; commentary: tomo II, 227–233).

9 The letter is in a manuscript in the Vatican Library (*Vat. Lat.* 2835, f. 11 *verso*). An unabridged transcription is provided in Ammirati 1983, 147 (*Appendice* I), and a partial one in Basile 1992, tomo II, 227–228.

manuscript there are about ten Latin epigrams composed for the *Beatricium*.[10] Basile argues that another epigram entitled *De quadam imagine marmorea* most probably also alludes to the same statue.[11] Subsequently, Tania Basile and Jean-Jacques Marchand were able to add a final sonnet by Tebaldeo for the *Beatricium*, which explicitly enunciates the *topos* of the challenge acted out between the artist and Nature.[12] More recently, Tebaldeo's poems relating to the marble portrait of Beatrice have been the object of a searching discussion by Lina Bolzoni.[13]

Antonio Tebaldi (Ferrara 1463–Rome 1537), who changed his name to Tebaldeo, was the preceptor of Isabella d'Este and moved with her to Mantua when she married Francesco II Gonzaga. This Ferrarese poet dedicated two more sonnets to Leone, and one of them (*Visto ho, Ambrosio gentil, l'opra tua rara*) is evidently in praise of *De Nola*, and indirectly an *elogium* of the city of Nola.[14] The close connection between Leone and Tebaldeo (the only poet involved in the *Beatricium* who is mentioned in *De Nola*), might easily suggest that Ambrogio was linked, first, to the court of Ferrara, and then of Mantua.

The most important discovery in this story was made by Pierre de Montera, who published a letter sent by Leone to Jacopo Sannazaro in 1493, in order to invite him to participate in the collection of poems, called "*Beatricium*", which he was compiling with the purpose of publishing it.[15] It was a bold request, because Sannazaro was one of the most renowned poets in Europe. Thanks to the wording in the letter we can deduce that Beatrice was still alive at this date. However, we have no evidence of the existence of Sannazaro's answer or of any poems by him dedicated to Beatrice, and perhaps we can assume that the sublime poet of *Arcadia* did not pay much attention to Leone's request.

Pierre de Montera was also able to find six epigrams by Ercole Strozzi (1473–1508), another poet of the Ferrarese court. Most recently Massimo Castoldi has continued to discover poems dedicated to Beatrice: two sonnets by Timoteo Bendedei, who was also Ferrarese, six Latin epigrams by Lancino

---

10   Rome, Vatican Library, *Vat. Lat.* 2835, ff. 11 *verso*-12 *verso*, 136 *recto*–136 *verso*, 202 *verso*–203 *verso*; in Basile 1992, tomo II, 227.

11   Modena, Biblioteca Estense, ms. alfa.T.9.18 (Lat. 681), f. 165 *recto*. On this manuscript see Basile, Marchand 1989, 63, n. Me[7].

12   The sonnet is *Sapi, lector, che quel legiadro aspetto*; in Marchand 1992, 872–873, n. 656; see also Basile 1992, 227.

13   Bolzoni 2008, 51–53, 157–167; Bolzoni 2010, 166–170.

14   This last sonnet is in Marchand 1992, tomo I, 425, n. 451; the other (*Se Marco dice che in quel clima vostro*) in Basile 1992, 378, n. 246.

15   Montera 1934.

Curti or Curzio († 1512), a man of letters from the Milanese *milieu*,[16] and a sonnet written by Giovan Francesco Caracciolo, the only Neapolitan contributor
to the *Beatricium* who has so far been found.[17] Castoldi has also shown how the
sonnet by Giovanni Pincaro *Qual sia qui sculpta odrai, se fermi il passo*, tradi-
tionally taken to refer to the bust of Beatrice d'Este sculpted by Gian Cristoforo
Romano (today in the Louvre),[18] could be interpreted more appropriately in
connection with the lost marble of Beatrice de' Notari.[19] Pincaro's verses were
formerly linked to Beatrice d'Este because he was active in the Ferrarese court;
however, we have seen that Leone had connections with this city, and consequently Pincaro's possible involvement in the *Beatricium* would not be out of
place. It would be tantalizing if to this already conspicuous collection of poems
one could add the nineteen Latin epigrams dated to the same years and written by the Milanese Giovanni Biffi (1464–1516), noted by Simone Albonico and
Giovanni Agosti.[20] These epigrams, published in 1493, refer to a marble statue
sculpted by "Malvico" portraying Beatrice, a Bolognese maiden, while, according to Leone, Beatrice was from Nola.[21] The correspondences between Biffi's
epigrams and the *Beatricium* are very evident and it is my conviction that these
verses are a response to Leone's poetic project. Nonetheless, the mistake about
Beatrice's origins has to be taken into consideration. It could be a simple typo-
graphical error, due to the fact that in the same incunable there are epigrams

---

16   For general information on Lancino Curti's *oeuvre* see Melfi 1985; about his library, Ganda
      1991; about Curti and the sculpture of his time and his tomb carved by Bambaia, Roth
      1980, and Kusch-Arnhold 2007; for Curti's stimulating connections with artists of his time,
      see Agosti 1990, 120–122, and Agosti 1998, 56–60.

17   The poems by Bendedei, Strozzi, and Caracciolo are published in Castoldi 1989; those by
      Curti in Castoldi 1992.

18   Starting from Bertoni 1919, 36 and 1921, 193 (in which only the first quartain of the sonnet
      is transcribed); see also Basile 1992, tomo II, 228. On the Louvre bust, see Bresc-Bautier
      2006, 189 (ML.10), and the recent catalogue entry by Julia L. Valelia, in Christiansen,
      Weppelmann 2011, 255–257, n. 101 (neither makes any reference to Pincaro's poem).

19   Castoldi 1992, 51–52, with the complete transcription. The first tercet is: "Beatrice che
      Nol'orna ancor viva / qui rapresenta e lei col corpo altrove / fruisse in terra il ben de
      un'alma viva"; Castoldi interprets it—in my opinion—correctly : "Beatrice, who adorns
      the city of Nola".

20   Albonico 1995, 91, makes a reference to the *Beatricium* and to Biffi's epigrams. These have
      been briefly discussed in an acute footnote by Agosti 1998, 58 (n. 52), in which he also
      mentions the existence of several verses by Antonio Plebano from Cremona dedicated to
      Beatrice marmorea that I have been unable to find.

21   "De statua marmorea a Malvico affabrefacta Beatrici puellae Bononiensi dicata"; Biffi
      1493, f. 10 *recto*.

related to a Bolognese context.[22] On the other hand, it is also possible that Biffi was not correctly informed and used the *Beatricium* only as an expedient or excuse merely in order to participate in a—possibly fashionable—poetic collection. The information about Beatrice's marble portrait, or the invitation to contribute verses in its praise, could have come to Biffi on a kind of 'disturbed connection', and the misinterpretation could have been caused by this faulty communication. In this reconstruction, the error is a fascinating piece of evidence of the extent of the fame of the *Beatricium* across the Northern Italian courts.

It is interesting to add here another, but very late response to the *Beatricium*. In his 1608 *Illustrium mulierum et illustrium litteris virorum elogia*, Giulio Cesare Capaccio mentions Beatrice at the very end of his collection of famous women, alluding to her peculiar death and to the poetic enterprise of the *Beatricium*. In addition, Capaccio also incorporates two Latin poems by contemporary author Carlo Pinto, devoted to the "Nolana Beatrix" and Malvito's marble portrait. It is quite surprising that at the beginning of the seventeenth century Neapolitan poets were still fascinated by the *Beatricium*, to the extent that they were producing poetic responses to it.[23]

The *Beatricium* constitutes what was at the time a unique example of a collection of poems dedicated to and inspired by a sculpture. Furthermore, it must be regarded as a highly significant precedent for the *Coryciana*, which is probably the best known poetic collection ever devoted to a sculpture, in this case Andrea Sansovino's *St. Anne with Mary and Child* commissioned by Johannes or Hans Goritz for his private altar in Sant'Agostino in Rome.[24] At the same time, the participations in the *Beatricium* allow us to re-trace the network of intellectual relations that Leone had interwoven in many different courts across the Italian peninsula. This network is extremely useful in visualizing geographically the cultural world of an Italian humanist.

The origins of the tradition of celebrating in poetry the speaking likeness and beauty of artistic portraits may be found in Petrarch's famed verses on the

---

22   At f. 6 *recto* there are verses dedicated to Giovanni Bentivoglio.

23   Capaccio 1608, 234–235. Capaccio's source is undoubtedly the *De Nola*. Carlo Pinto is a little known author, who published Pinto 1607 and the chronicle of the fire of Montevergine, Pinto 1611, both printed by Giovan Giacomo Carlino. I also found mention of a collection of epigrams for the death of the Viceroy Count of Lemos, but I was not able to find a copy of it (Pinto 1602; mentioned in Toppi 1678, 314).

24   The bibliography on the *Coryciana* is considerable. As a reference see the critical edition by Ijsewijn 1997; see also Bober 1977; Haig Gaisser 1995; Perini 1997–1998; Nagel 2000, 104–105, 248–249; Sodano 2001; and Rijser 2008. On poetic anthologies in the Italian Cinquecento see Bianco, Strada (eds.) 2001.

portrait of Laura supposedly painted by Simone Martini (*Canzoniere*, LXXVII–LVXXVIII).[25] It is also true that the flourishing combination of word and image—in particular poetry and portraits—was explicitly explored in the design and production of medals, another genre extremely *à la page* at that time. One only needs to recall the renowned medal, cast by Gian Crostoforo Romano, portraying Isabella d'Este on the *recto* and bearing on the reverse a motto especially created by the humanist Niccolò da Correggio in 1498; a version of the medal cast in gold was donated by Isabella to Serafino de Ciminelli, who answered with a sonnet thanking her.[26] Furthermore, it is well known that as early as the summer of 1491, Isabella was eager to have a marble bust carved by Romano, which apparently was indeed executed by the sculptor, since it was celebrated in verses by Tebaldeo probably around 1497.[27]

The *Beatricium* poems follow the classical *topoi* of the dialogue between Art and Nature: sculpture which can attain an incredible level of mimesis to the point of confusing the spectator, or the work of art which lacks only the power of speech in being lifelike. When we read all the poems, we realize that none provides descriptive details of the statue, although "Malvico" is often named, and it is difficult to understand what exactly the sculpture looked like. We can only infer that it was in marble, and was carved before 1493, since it is cited in Leone's letter to Sannazaro.[28] One would think that none of these poets had seen with his own eyes the sculpture, even if Pierre de Montera suggests that Tebaldeo could have seen the *Beatrice marmorea* when he visited Naples in 1493.[29] During this period we know that Tommaso Malvito was in Naples. He moved to the capital of the Kingdom almost certainly in the early seventies, after which he followed Francesco Laurana in order to work at the Saint Lazare Chapel in the Cathedral of Marseille [figs. 54, 55, 56], where he is documented as being from at least January 1479 until the spring of 1483,[30] but before February 1484 he had returned to Naples.[31]

25   Collareta 1988–1989; Bolzoni 2008, 10–14, 75–81; Bolzoni 2010, 153–157. See also Pignatti 2008.

26   On Romano's medal see Syson 1997; Agosti 2005, 124, 146 (footnote 78), 477 (footnote 48); Brown and Lorenzoni 2007.

27   On Tebaldeo's sonnet *Firmar non te potevi in loco dove* see Basile 1992, n. 251 (commentary: tomo II.2, 249–250); Brown and Lorenzoni 2007, see also de Castris 2010, 15–16.

28   Agosti 1998, 58, notes that the publication in 1493 of Biffi's epigrams for the *Beatricium* is a proof that the sculpture was carved before this year.

29   Montera 1934, 195, footnote 5.

30   Barthélemy 1884, 633–635, 637; and also Kruft 1995, 194–203.

31   Filangieri 1883–1891, III, 82.

In the light of Malvito's collaboration with Laurana, we can try to visual-
ize the marble portrait of Beatrice de' Notari, which was in all probability a
bust. Francesco Laurana also worked between Naples and Sicily and is well
known for his marvelously refined and serene bust portraits, such as the one
of Beatrice d'Aragona (New York, The Frick Collection) [fig. 52], or of Eleonora
d'Aragona (Palermo, Museo Regionale di Palazzo Abatellis). A further argument
in support of its being a bust is the singularity, to say the least, of the idea that
Leone commissioned a full-length marble statue of a young woman who was
still alive; this would have been both very expensive as well as seeming preten-
tiously excessive. We have no information on Ambrogio's economic means, but
we can imagine they were not limitless, while Beatrice was born to a provincial
noble family, and had certainly no royal blood. Conversely, the bust shape ac-
cords with contemporary Neapolitan taste, would have been more affordable
for Leone, and is perfectly appropriate for a poetic celebration. Following this
line of argument, Damianaki has argued—even if in a cautious and tentative
way—that the Louvre female portrait generally ascribed to Laurana (from
Wilhelm von Bode to Hanno-Walter Kruft) could be the Beatrice by Malvito
[fig. 53].[32] Damianaki wants to argue that the reason for this re-attribution lies
in the more modest artistic quality of the Parisian bust compared with the
other autograph works by Laurana. Although, in my opinion, this argument
raises great difficulties given the stylistic closeness between the Paris bust and
the works by Laurana and his workshop in Marseille, it is plausible to think
that the bust owned by Leone was a very similar object to the Louvre marble.

The Beatrice de' Notari portrait fits perfectly within the Italian Renaissance
court culture and its taste.[33] The ambiguity which has led to the Pincaro sonnet
being connected with Gian Cristoforo Romano's bust of Beatrice d'Este [fig. 57]
is almost an inverted demonstration that these two works of art, depicting two
distinct Beatrices, share the same cultural environment. It is well known that
the sculptor Gian Cristoforo Romano was also an advisor to Isabella d'Este.[34]

---

32   Damianaki 1998, 368, n. 14; 2000, 92–96; lastly 2008, 186–193, n. 6. For the traditional at-
      tribution to Laurana see Bode 1888, 216–218, and Kruft 1995, 153–156; the complete bibli-
      ography on the bust is in Bresc-Bautier 2006, 156, (MR.2597).

33   For a general panorama, see Bolzoni 2008; Bolzoni 2010; Christiansen, Weppelmann 2011.

34   On Gian Cristoforo Romano and Isabella d'Este the correspondence published by Brown
      and Lorenzoni 2007 is fundamental; for a selected bibliography see Brown, Lorenzoni and
      Hickson 2002; Campbell 2004, 87–113 (Chapter 3: "Isabella's Cupidity: Collecting and Literary
      Production"); Pidatella 2006; and also Francesco Caglioti in Beltramini, Gasparotto, Tura
      2013, 194–196. A comprehensive and coherent monograph on Gian Cristoforo Romano, a
      key figure of the Italian Renaissance and its antiquarian culture, is still waiting to be writ-
      ten; for a general overview of his career Venturi 1888 remains a point of reference; for more

Romano was responsible for procuring outstanding works of art for her, for which she showed an eager appetite. For example, in July 1505 Gian Cristoforo suggested that she purchase a silver inkwell made by Caradosso Foppa (1452–1526 or 1527), as splendid as it was expensive, which had been originally made for the Cardinal Giovanni d'Aragona.[35] It happens that both this "calamaro" and Caradosso himself have close links to Ambrogio Leone, and these connections will be explored in the following section of this essay.

2 Caradosso's Inkwell

Caradosso is one of the few artists cited by Leone. In his treatise *De nobilitate rerum* (1525, ch. 41) Leone mentions two—in his opinion—excellent Milanese sculptors of his time: one of them is Daniele Arcioni, a glass-artisan now almost unknown who worked at "niello" and fabricated a beautiful salt-cellar, and the other is the goldsmith Caradosso.[36] After comparing Caradosso to Praxiteles or Lysippus, Leone provides an exact description of one beautiful piece by him, which was an inkwell adorned with four plaquettes (the Rape of Ganymede, the Battle of Centaurs and Lapiths, Hercules and Cacus, Hercules and the Nemean Leon) and assembled for the Cardinal Giovanni d'Aragona [figs. 58, 59, 60, 61]. Leone was even aware of the price of the inkwell: 1,500 *aurei*.[37] Leone significantly concludes his digression with an assertion of the

---

recent studies see Ceriana 1999; Leone de Castris 2010, 3–63 (although in this last book the considerations on Romano's Neapolitan period are questionable); Pidatella 2011; and Pidatella (forthcoming); new ideas on Romano as a sculptor were proposed by Francesco Caglioti in Christiansen, Weppelmann 2011, 302–306.

35 Gian Cristoforo Romano to Isabella d'Este; Milan, 20 July 1505: "Ho receuta da Vostra Signoria la risposta del vaso [on which we will speak further] et òlla mostrata a Charadosso, el quale cerca ogni via per acelerarse e venire im persona con el vaso a trovar Vostra Signoria [...]. Se altro acascarà di bello, ne advisarò Vostra Signoria: el ditto Caradosso ha il più bel calamaro che sia a l'età nostra, quale altre volte el fece quando lui stette col reverendissimo Cardinale di Ragona, e lo vorria vendere, ma ne domanda milli ducati, e veramente se Vostra Signoria si trovassi diecimila ducati in cassa io vi exortaria a non lo lassare, perché l'è cosa unica [...]"; in Venturi 1888, 112–113; and more recently in Brown and Hickson 1997, 22.

36 Morelli (1800, 204–207, footnote 119) had already studied this text, which has been discussed at length by Brown and Hickson 1997, 9–10, 12–14, 20–22. The mention of Daniele Arcioni was already noted by Malaguzzi Valeri 1917, 315, who incorporated Arcioni among the artists at the court of Ludovico il Moro.

37 Leone 1525, ch. 41, English translation by Brown, Hickson 1997, 9–10: "Indeed in our time the subtle and ingenious works of two distinguished Milanese sculptors are highly

primacy of sculpture over painting, justifying this in the light of its material significance. Thus, in order to support his assertion, Leone does not use monumental marble sculpture as a proof, but two small precious objects (a salt-cellar and an inkwell) instead. This choice is particularly relevant because it seems to counter the trend in the context of the High Renaissance, in which Michelangelo's giants had already come into being. However, it is unnecessary to recall how acclaimed these exquisite and unique small works of art were, from Erasmus, who owned a collection of medals and plaquettes, to Pietro Bembo, whose cardinal's seal, for example, is an astonishing piece of Italian High Renaissance sculpture the creator of which remains unidentified.[38]

----

regarded and talked about. The one is named Caradosso and the other Daniel Arcioni. Arcioni especially excels in the genre they call by the neologism 'niello'. He is no less brilliant in [working] the glass with the founders themselves call enamel. In these areas he has demonstrated such subtlety and proportion that he held in the highest honor by all who excel in this art. His work, moreover, [benefits] from such arrangement, mounting and grace that it challenges the most admirable images of antiquity. Among [his works] there is a salt-cellar known to have been made from fourteen ounces of silver. He sold it to someone in Rome for seven hundred gold pieces of aurei. To be sure Caradosso is better known for the eminent images and sculptures he produced. He was considered superior to other foreign and Italian artists. Because of the superiority of his artistic skill or natural talent, even the experts hardly dare indicate whether the works he produced were created by Praxiteles, Lysippus or a more recent artist. There is still extant an inkwell of his—to leave aside other splendid and well-known works—in which all the skills of his art are so present that there seems to be no room left for improvement. On one side of it are seen nude men mounted on horses coming to the aid of some boy whom an eagle has snatched up into the sky. Indeed these men watching the bird carrying off the boy show such courage and resolution that they also seem to want to take off with their horses. In these figures Caradosso shows the rape of Ganymede. On the second side is the battle of the Centaurs with the Lapiths. On the third side is Hercules [holding] Cacus down [with one knee] and pressing on his throat with his left hand [while he] kicks him violently in the stomach with his other knee and forces his mouth shut with [his right] hand. On the fourth is seen Hercules subduing the lion. [It is] so beautifully carved that you can feel the man stirred by anger and [hear] the lion groaning in pain. Because of the subtlety and excellence of the work, many sulphur tablets have been spread [far and wide] throughout Italy. Consequently his work is held in high regard throughout the whole of Italy. Giovanni d'Aragona, the son of Ferdinando, is said to have offered 1, 500 gold pieces for it. May you accept these things as an explanation of the great genius of the sculptor's art. We think that these works make it obvious which of these arts is the more perfect. Sculpture is the more perfect both because of the material it employs and its innate greatness".

38   On Erasmus's collection see Lewis 2008. On Bembo and the visual arts see the impressive exhibition catalogue Beltramini, Gasparotto, Tura 2013 (in particular on the sigillo, the

The existence of Caradosso's inkwell as described by Leone is confirmed in *De Cardinalatu* by Paolo Cortesi, who mentions having actually seen it in the Roman palace of Cardinal Giovanni d'Aragona.[39] D'Aragona died in 1485 and therefore one has to deduce that the *calamaio* was commissioned before this date. On account of their iconographic correspondences, three surviving bronze plaquettes have been connected with the inkwell described by Leone, and have been considered works after Caradosso.[40] The significant question here is if Leone actually saw the inkwell, and if he did, where.

The lengthy encomium of Caradosso written by Leone is one of the most important testimonies for the activity of this still today obscure Milanese goldsmith.[41] In this essay there is no space to expound the theories on his

---

entry by Davide Gasparotto and Alessandra Guerrini, 369–370), as well as the volume Beltramini, Burns, Gasparotto 2013; and Gasparotto 2014 (in particular on the seal, 133).

39   "Qualis [officina pictorum] nobis pueris ea videri potuit quae in Ioannis Aragonis Senatoris domo erudita fuit, in qua a Caradoxo Mediolanense anaglyptum atramentarium antiquo opere ex argento caelari vidimus" ([The workshop for painters] such as we saw as a child in the cultivated household of Cardinal Giovanni d'Aragona, where we saw an inkstand cast in silver with relief work done in the antique fashion by Caradosso of Milan); Cortesi 1510, G.iiiir; in Weil-Garris and D'Amico 1980, 84–85. Moreover, fra Sabba da Castiglione (begun ca. 1505; ed. cited 1569, f. 115 *recto*): "Et chi con le gentilissime opere del mio Caradosso, il quale, oltra la cognition grande delle gioie, in lavorar di metallo, in oro et in argento, o di tutto o di basso rilievo, all'età nostra è stato senza paro, come si può vedere nella città di Milano per un suo calimaro d'argento di basso rilievo, fatica d'anni vintisei, ma certo divina". See Brown and Hickson 1997, 31, for whom the "twenty-six year gestation needs to be clarified". On Sabba da Castiglione see the recent book ed. by Gentilini 2006.

40   The first connection between Leone's *De nobilitate* and the *Battle of the Centaurs* plaquette was elaborated by Muntz 1883. However, see, as essential references, Brown and Hickson 1997, 21, and the very recent and detailed entry, with the comprehensive bibliography, in Rossi 2011, 147–149 (no. IV.12). The specimens cited by the authors are in the following museums: *Rape of Ganymede* (cm 4, 9 × 4, 9), New York, The Metropolitan Museum of Art; *Battle of Centaurs and Lapiths* (cm 4, 8 × 4, 9), Washington DC, National Gallery of Art; *Hercules and Cacus* (cm 6, 08 × 3, 65), Washington DC, National Gallery of Art. As Brown and Hickson point out, the original plaquettes have been dated by scholars to the late 1480s and the early 1490s, a chronology which seems incompatible with the death of Giovanni d'Aragona (1485). In order to find a solution for this problem, Brown and Hickson suggest it could be due to a merging in the literature of several inkwells into a single one and/or to a confusion by Leone.

41   Caradosso's ability was praised also in Gaurico's *De sculptura* of 1504 (ed. Chastel, Klein 1969, 263), by Vasari (1550, 531; 1568, I, 503) and Cellini (ed. Bacci) 1901, I, xxvi, xliii, liv; and Cellini (ed. Milanesi) 1857, 30–31, 72–5, 89–90, 95–6), and in *Trattato della pittura* (1584) by Lomazzo (ed. Ciardi 1973–1975, I, 325; II, 287, 534, 549, 550). Nevertheless, Leone's

personality and artistic production. However, alongside the documentary studies carried out by Clifford Brown, there is a considerable but still problematic scholarly tradition (beginning with Ulrich Middeldorf) that identifies Caradosso with the goldsmith known as Moderno, whose plaquettes were most probably collected also by Erasmus.[42]

In the letter from Gian Cristoforo Romano quoted above, the central piece of information regards the possible purchase by Isabella d'Este of a crystal vase mounted in silver which belonged to Caradosso.[43] While it is not totally clear whether the vase was made by Caradosso or just owned by him, we do know that Gian Cristoforo was responsible for the difficult negotiation.[44] Caradosso was invited to hasten to Mantua in order to show the work personally to Isabella.[45] Having to pay the customs duty for crossing the border between the Duchy of Milan and the Gonzaga territories, Caradosso found a cunning way round it: since Niccolò Orsini Count of Pitigliano had the right to free passage, Caradosso decided to wait for an "homo del conte de Pitigliano" with whom he could cross the border without paying, pretending that the vase belonged to the Count.[46]

---

description of the Aragona inkwell is without doubt the most precise we have of a work by Caradosso. For his literary fortune see Brown and Hickson 1997, 37; and also the online biography by Collareta 2010.

42   Brown and Hickson 1997, and Brown and Lorenzoni 2001. On Erasmus and the visual arts the homonymous article by Panofsky 1969 remains an essential reference; in particular on Erasmus' collection see Lewis 2008. A long *querelle* has been developing around the identification of Caradosso with Moderno. Agosti 1990, 131–132, footnote 87, reported that it was an idea of Middeldorf and he announced a study, developing this idea, by Collareta. Collareta has later suggested this overlap of the two artists in Romano and Salsi 2005, 156, n. 11.21, as well as in several earlier lectures. At the same time Venturelli 2002, 145–157 has also argued in favour of Middeldorf's hypothesis. Brown 1997, 65, and recently Rossi 2011, 145 and 165–166, are opposed to this identification, and supports the prevailing idea (firstly by Bode 1904, reinforced by Lewis 1989) according to which Moderno was the Veronese Galeazzo Mondella. Gasparotto 2008, 90–91 seems not to agree with either of these solutions. My own view of the matter is closer to this last position.

43   See footnote 35.

44   Brown, Hickson 1997, 20, notice that nowhere in the documents is it affirmed that the jar was executed by Caradosso. However, the *lectio facilior* remains that he was responsible for the mounting.

45   See *supra* the letter dated 20 July 1505, footnote 23.

46   Gian Cristoforo Romano to Isabella d'Este, Milan, 9 August 1505: "Circa al fatto di Caradosso, io ebi la littera e ogni giorno l'ò solecitato, ma lui aspettava un homo del Conte de Pitigliano [Niccolò Orsini], el quale à licentia posser portar senza datio le robe del ditto conte, e con lui voleva acompagnarse con dir che 'l ditto vaso era cosa del conte, e subito

This anecdote brings to light a relationship between Caradosso and Niccolò Orsini, who was count of Pitigliano (Tuscany), and of Nola as well. A portrait medal of Niccolò has in fact been attributed to Caradosso [fig. 62].[47] In this, Orsini is depicted as *condottiero* of both the papal and the Florentine troops, and, since he entered the service of the Venetians in 1496, it is possible to date the medal to between 1485 and 1495. As we will see below, Orsini would become the hero of the resistance of the Serenissima against the League of Cambrai. It is impossible that Leone, who was firmly rooted in the Venetian cultural *milieu*, could not have known the Count of Nola, who was a key figure in Venetian politics from 1496 to his death in 1510.[48] It is plausible to conjecture that all three, Caradosso, Orsini and Leone, were very likely connected to each other.

3   Tracing Interconnections: *De Nola*, Girolamo Mocetto, Niccolò Orsini, and the League of Cambrai

Niccolò Orsini was the second to last count of Nola. The last was his grandson Enrico Orsini; it is to Enrico that Leone's *De Nola* and the posthumous edition of *De nobilitate rerum* are dedicated.[49] By considering the relationship between

---

ch'el venga lui, venirà via; se non, se pigliarà altra via. Basta che più presto che si potrà Vostra Signoria lo vederà, qual son certo che piacerà a quella"; in Venturi 1888, 114; and more recently in Brown, Hickson 1997, 22–23.

47   The attribution was suggested by Foville 1911, and followed by Hill (1930, 172; and 1931, 100, n. 196); Middeldorf, Goetz 1944, 11, n. 71; Hill, Pollard 1967, 39, n. 196; *Glorious Horsemen* 1981, n. 49; Wilson 1983, 98, n. 7; Pollard 2007, I, n. 221, with no reference to Gian Cristoforo Romano's letter. Brown and Hickson 1997, 23 (footnote 3) cite the attribution, but they do not discuss it. The medal (approximately 41 millimetres in diameter) depicts on the obverse the profile of Orsini, facing to the left, in armour, and around: "Nic(olaus) Urs(inis) Pet(iliani) et Nol(ae) comes Sante Rom(anae) Eccle(sie) armor(um) cap(itaneus)"; and on the reverse Orsini riding to the right accompanied by two halberdiers, and around: "Nic(olaus) Urs(inus) Petiliani et Nolae comes Reip(ublicae) Flor(entiae) cap(itaneus)". Some specimens are in: Washington, The National Gallery of Art (1957.14.789); Florence, Museo Nazionale del Bargello (n. inv. 6112; 6113; 6114); London, British Museum (1969, 0111.96); London, Victoria and Albert Museum (4565–1857); New York, American Numismatic Society (1949.145.1).

48   Even if the *De nobilitate rerum* was published posthumously in 1525, it is possible to date its gestation between Ambrogio's last years in Nola and first years in Venice, as demonstrated in de Divitiis, Miletti 2016.

49   The dedication of *De nobilitate rerum* was written by Ambrogio's son, Camillo, but it is significantly addressed to Enrico Orsini. Perhaps Camillo respected his father's intentions. On the relationship between Leone and Enrico Orsini, see Miletti in this volume, 11–15.

the Orsinis, Leone and the critical moment in which *De Nola* was published (1514), we will open the last section of this essay, whose aim is to show how relevant these interconnections are not only to the analysis of Leone's *oeuvre*, but also to the understanding of literary and artistic production in the context of the extremely dramatic war of Venice against the League of Cambrai.

Niccolò Orsini, like every *condottiero* of the time, was accustomed to changing factions and to serving different Italian states, which were continually at war with each other during the Renaissance. It follows that he had commissioned two very similar sepulchral monuments, the first one, definitely for Orsini, is today in Santo Stefano in Fiano Romano [fig. 66], for which it was probably originally intended, and later a second one in the church of Santa Maria delle Grazie in Ghedi, now preserved in the Museo di Santa Giulia in Brescia [fig. 65]. A third [fig. 64], as we shall see, would be built in his memory after his death by the Republic of the Serenissima in the Venetian basilica of Santi Giovanni e Paolo, or San Zanipolo, as it is called in local dialect.[50] Nonetheless Orsini obviously remained very close to his important County of Nola. When his wife Elena dei Conti died in 1504, she was buried in the convent of San Francesco (today San Biagio) in Nola,[51] which had been rebuilt as a family church by Niccolò I Orsini in about 1372 in order also to house an Orsini pantheon.[52]

---

50 Pompeo Litta (1819–1883), IV, *Orsini di Roma*, tavola XI) already noticed Niccolò Orsini's three tombs. For more details see Markham Schulz 1987, 321, footnote 50; and also Fattorini 2011 (in particular footnote 14; although with no reference to Schulz). A drawing of the tomb in Fiano Romano is also included in Cassiano dal Pozzo's *Museo Cartaceo* (The Royal Library at Windsor Castle, cod. 201 Albani, fol. 194, inv. n. 11.908).

51 Bruscalupi 1906, 278, and 640–641 footnote 107, in which the lost epitaph is transcribed: "Nicolaus Ursinus Nolae Pitilianique comes dilectae coniugis Helene Conti hic corpus voluit humari, et pro eius anima semel in ebdomada missam et semel in anno anniversarium celebrari, ob quod Conventui huic in palmarum planitiae telluris iugera quinquaginta donavit. A. D. MDIV. VI Idus Iunii [8 June]".

52 The church has undergone much damage and restoration. The only memory of the Orsini tombs is the empty sarcophagus, which still remains in the sacristy. Leone (1514, bk. III, ch. 13), writing about the church, did not mention the Orsini funerary monuments. The earliest and most interesting description we know of them is provided by Guadagni in 1688 (ed. Toscano 1991, 210–211). Subsequently Remondini (1747–1757, I, 206–207) described the dismantled monument in the sacristy. On the sarcophagus see also Avella 1990, 90, footnote 38; Avella 1996–1999, II, 334, figg. 629–631; Bock 2001, 455–456; Toscano G. 1989; Toscano G. 1996. I have presented two papers on this topic with some new considerations (Renaissance Society of America, Washington D.C., 22–24 March 2012; London, The Warburg Institute, 16 November 2012), and I am currently preparing an article on the subject.

In 1508 the allies of Cambrai attacked the Republic of Venice. It was a completely unequal conflict, since some of the most powerful European states took part in the League, promoted by Julius II, such as the Holy Roman Empire, France, Naples, Ferrara and Mantua. A Venetian defeat was almost inevitable. At that time Niccolò Orsini and Bartolomeo d'Alviano were the *capitani* of the Serenissima. The outcome of the battle of Agnadello (May 1509) was disastrous: several generals escaped and Alviano was taken prisoner. After the loss of Padua, following a heroic resistance, the destiny of Venice seemed tragically marked. Thanks to the Serenissima's diplomacy—the Venetians even threatened to ally themselves with the Ottoman Empire—as well as Orsini's prudence and fidelity, in July 1509 the wind began to change: Andrea Gritti, Orsini's *provveditore generale*, was able to reconquer Padua. A few months after, in January 1510, Orsini died in Lonigo. The war had not ended, and although the situation was improving for the Venetians, it was still extremely risky.[53] The Serenissima was in need of heroes to celebrate on the altar of the Fatherland, and Orsini was a perfect candidate. Giovan Battista Egnazio, a close friend of Leone, composed both a *oratio funebris* and the epitaph for Niccolò, and it is very likely that Ambrogio attended the funeral of the Count of his hometown Nola.[54] A sepulchral monument was immediately commissioned and is still today in the right transept of Santi Giovanni e Paolo. It has been convincingly attributed to Antonio Minello by Anne Markham Schulz, who adds that the far from excellent quality of the tomb can be explained by the economic crisis due to the war.[55]

Some years later, around 1515, when the war was finally coming to an end,[56] the *procuratore* of the convent of Santi Giovanni e Paolo, Giorgio Emo, hurried along the completion of the massive stained glass window which was to adorn

---

53   For a *résumé* of the facts and the most important historical sources see Lenci 2002.
54   Egnazio 1510. See Zambelli 2007, 260; and also Miletti in this volume, 11–18. The epitaph reads: Nicolao Ursino Nolae Petilianique Principi longe cla|rissimo, Senensium, Florentini Pop[uli], Pii, Sixti, Innocentii, | Alexandri Pontificum Maximorum, Ferdinandi Alphonsique iunioris re[gum] | Neapolitanorum imperatori felicissimo, Venetae demum Rei P[ublicae] | per xv annos magnis clarissimisque rebus gestis, novissime | a gravissima omnium obsidione Patavio conservata, virtutis et | fidei singularis, S[enatus] V[enetus] m[onumentum] h[oc] p[ublice] p[osuit]. Obiit aetatis anno LXVIII MDIX.
55   Markham Schulz 1987.
56   The conflict stopped at the end of 1515, but was already petering out in the months before this. The peace was signed in Brussels in 1516. The most important political change in the scenario had been the accession to the throne (January 1515) of François I de Valois, who was very interested in the Duchy of Milan and wished to reconquer it relying crucially on Venetian support. Maximilian I gave back the *terraferma* to the Serenissima only in 1517.

the right transept, a huge work of art that is quite unique in the Venetian context [fig. 67, 68].[57] Giorgio Emo was appointed *procuratore* in 1513, but before
that he had been in charge of both the sumptuous funerals in San Zanipolo
held for Niccolò Orsini (1510) and Leonardo da Prato (1511), heroes of the war
against the League of Cambrai. Lionello Puppi has noted the strong connection between the War of the League of Cambrai and the iconography of the
vetrata, in which at the first level four warrior saints are depicted.[58] Puppi's
attractive interpretation suggests that the armed saints are intended to allude
to the *condottieri* who had defended and saved Venice in those years and who
were all buried in San Zanipolo.[59] According to Marin Sanudo, under Giorgio
Emo's rule as *procuratore* the monuments (the three carved by Minello) of the
three heroes Niccolò Orsini, Leonardo da Prato and Dionigi Naldi were un-
veiled during a special ceremony on December 24, 1514, even though they were
not definitively finished.[60] The tombs were completed in the following winter
of 1515, significantly at the same time that the window also was completed. The
interconnections between Giorgio Emo, the tombs of the *condottieri* and the
decoration of the stained glass window emerge clearly here.

The artist who executed part of the lower section of the window was
Girolamo Mocetto. His signature, "Hieronimus Mocetus faciebat", is still clearly visible on the base of the figure of Saint George [fig. 68].[61] Mocetto was a

---

57   Marin Sanudo in 1515 wrote "[Giorgio Emo fa] compir la fanestra grande di coro", and
     Romano 1985, 85–86 persuasively argued that the gigantic window was concluded in 1515
     or a few months later. Most probably, the decoration of the upper part of the window was
     begun around 1473, and the completion of the lower part was commissioned at least by
     1510 (according to a document found by Lionello Puppi and published by Romano 1982,
     53).

58   Puppi 1982, 33. Saint George and Saint Theodore are represented at the sides; the ico-
     nography of those in the center instead is not clear. Puppi proposes an interpretation of
     them as Saints John and Paul, the two brothers, martyrs and Roman soldiers, to whom the
     basilica is dedicated, undoubtedly the most realistic and straightforward solution.

59   This fascinating and in my opinion correct interpretation has been followed and devel-
     oped by Romano 1985, 85; and De Vito 1986. For the relations between the War of the
     League of Cambrai and Venetian contemporary arts, see also: Howard 1985; and Fontana
     2009–2010.

60   See Romano 1982, 66. For the tombs of Leonardo da Prato and Dionigi Naldi, see Markham
     Schulz 1987.

61   The question of the authorship of the different parts of the *vetrata* remains a conun-
     drum that I do not intend to discuss here. Taking for granted the considerable presence
     of Mocetto, who very visibly signed one of the panels, scholars are divided among those
     who suggest that the window is the result of a heterogeneous collaboration (Romano
     1981; Romano 1982); those who defend the traditional attribution to Bartolomeo Vivarini

painter and, being born in Murano, his initial apprenticeship was dedicated to the art of glass-making.[62] He is better known as an engraver, a disciple of Mantegna who working with prints spread many of the great Paduan artist's ideas, at times freely interpreting them.[63]

The most interesting point in this context is that Mocetto was also in charge of the execution of the four engravings included in *De Nola*. Only the plate of *Ager Nolanus* is signed at the bottom left: "HIE. MOC." [figs. 1, 5, 6], but in the *prefatio* Leone names Mocetto as the artist who assisted him in the making of the four engravings.[64] Serena Romano, in a suggestive hypothesis, sees a link between Mocetto and Leone in the Orsini family and in the closeness—physical but also symbolic—of the window to Niccolò's tomb.[65] As the present volume strongly underlines in the context of a comparative analysis, Leone was a Nolan with deep roots in Venice and most probably in close dialogue with Venetian literati such as Giovan Battista Egnazio, Marcantonio Michiel, Marco Musuro and Aldo Manuzio, but who still retained strong connections with his hometown in Campania.[66] Moreover, he had built up an impressive network

---

(Sponza 1982; Sponza 1996, 105–106); and those who have more recently proposed an attribution to the painter Giovanni di Niccolò Mansueti (Fossaluzza 2012).

62   The main study on Mocetto is the monograph by Serena Romano 1985: for some hypotheses on his birth see *ivi*, 25–26. After the work for San Zanipolo, Mocetto reappears only in 1531, the year of his will and of his death (*ivi*, 98). A cycle of frescos for the façade of the Palazzo al Ponte dell'Acqua Morta in Verona has been attributed to him and its chronology can certainly be placed between 1517 and 1521. The principal scenes depict the Continence of Scipio and the Justice of Trajan; they are in a very poor state of conservation and are today preserved in the Museo di Castelvecchio. The attribution to Mocetto was proposed in the 19th century and has often been discussed. Serena Romano (1985, 92, 98 e 106) summarizes the different opinions and, even though she does not totally agree with the ascription, includes the Acqua Morta cycle in either Mocetto's or his workshop's production. These frescos are interesting in the context we are focusing on in this essay, since they are related to the war of the League of Cambrai and to the re-possession of Verona by the Venetians. According to Romano and to the even more accurate studies by De Vito (1986), the cycle was executed to celebrate this event. If the reference to Mocetto's ambience is correct, we should point out that he was involved in two projects of decoration linked to the development of the war and financed by the Serenissima. An analogous case is that of Antonio Minello, who sculpted the *condottieri* tombs in San Zanipolo.

63   On Mocetto's mantegnesque vein, see Agosti 1994; and van der Sman 2003.

64   See also Lenzo in this volume 76–80.

65   Romano 1985, 92, implies that this is the only possible explanation for the involvement of Mocetto in a book about Nola, a "far-off Southern city".

66   On the relations between Leone and the intellectual milieu in Venice see de Divitiis, Lenzo and Miletti in this volume, 1–10.

of relations with several courts across Italy, and beyond. We do not know how Leone and Mocetto first came into contact, but their relations must have originated in the common intellectual friendships within the Venetian *milieu* and are most probably datable to a period prior to the commission of the stained glass window. In 1514 *De Nola* had already been published: the complicated preparation of the plates must therefore have been done previously. As we can read in the *prefatio*, Leone attached great importance to the *De Nola* plates and had personally entrusted them to Mocetto, whose name is explicitly mentioned. Ambrogio undoubtedly knew that he was one of the most *à la page* engravers of his time and this collaboration is an additional demonstration of Leone's artistic sensibility and the attention he paid to contemporary artists. If the idea advocated by Serena Romano (that Leone met Mocetto thanks to the Orsini tomb commission) cannot be totally accepted, the protagonists of this triangle of connections could be seen to be related in another way: it might have been someone in Leone's *entourage* that proposed Mocetto to design the *vetrata* of San Zanipolo. On the other hand, it is important to remember that a connection between Mocetto and the Dominicans of Santi Giovanni e Paolo could date back to the production of the engraving of the Calumny of Apelles, based on a drawing by Mantegna, which is depicted taking place in the *campo* di San Zanipolo, the large square adjacent to the church.

We can assume that, during the really critical years of the war against the League of Cambrai, Leone was preparing, in Venice, his book "*de Nola urbe patria nostra*", as he affectionately calls his volume in his well-known letter to Erasmus.[67] The freedom of Venice was seriously in danger, and it was perhaps not the best of times for bringing out this kind of book. Nevertheless, one could suggest a different reading of the timing of this publication. As we have seen, Orsini had been the first hero of the war, and, although he did not die in battle, his memory was chosen to embody Venetian resistance. Orsini was the count of Nola and Leone's *opusculum* could have been duly appreciated as part of the rhetoric of the wounded fatherland. In a passage of *De Nola* (I, 13), in which Niccolò Orsini appears as well, Nolans are said to be distinguished for their loyalty. Furthermore the book is dedicated to Niccolò's heir. Praise of the feudal city would have been easily interpreted as praise for Orsini's origins. We have no information about the gestation of *De Nola* though it probably took a

---

67   Leone's letter to Erasmus (dated July 19, 1518) and the response are published in Erasmus (ed. Allen) 1913, 352–355. On Leone and Erasmus see Miletti 2016, 354–355; and Spruit 1960; Vecce 2000; Zambelli 2007, 259 footnote 20. On Erasmus in Venice see also the volume edited by Olivieri 1995.

long time.[68] Even if the idea and the composition of the book as a whole cannot be linked to the war against the League of Cambrai, this episode perhaps constituted a further incentive for its publication.

In this essay I have sought to show the dense network of relationships which Ambrogio Leone built up over the course of his life. Leone's artistic world arises naturally from these relationships. If we follow the interconnections, contacts and acquaintanceships which make up this network we draw the artistic map—so to speak—of a typical Italian humanist. The present study is merely a starting point for a further fruitful exploration of this map in order to discover other encounters and imagine other perspectives. *De Nola* is a cornerstone of Renaissance antiquarian studies. The ambitions which lay behind the unfinished *Beatricium* were of the highest; it was a literary undertaking often only referred to in passing but its extent and its resonance throughout the whole of Italy need to be brought out in order to understand Leone's aims which were far from being provincial or courtly. Leone was aware of artists working in most of Italy, from Malvito in Naples to Mocetto in Venice, and in the courts of Ferrara, Mantua and Milan. We therefore need to see Leone as a humanist of his time, whose knowledge of science, literature and the visual arts engaged fully and consciously with the whole Italian scene.

---

68   We only know that it was written for the most part in 1512, as Leone (1514, bk. II, ch. 1, f. xxii *recto*) says. See Miletti in this volume, 34–40, 42–44, who demonstrates that many quotations present in the *De Nola* are related to books published by Aldo Manuzio between 1512 and 1514, in other words shortly before the publication of *De Nola*.

CHAPTER 6

A Civic Duty: The Construction of the Nolan Memory

Giuliana Vitale

In preparing to write the history of his native city, Leone declares that he has not found any previous work on the subject; his aim therefore, in his own study, is to give future readers in Nola the possibility of comparing their own present with the city's past, at least as it is told in the account of it he has been able to reconstruct in his own time.[1] They would thus be able to use the information he has gathered through his own experience and the testimonies of the elderly which he has recorded.[2]

But Leone's history of the city, as he states in the preface to the book, had a deeper, more solemn purpose than merely being, as these remarks might suggest, of informational value: his description of the essential values expressed in the life of the city would be of use both to its inhabitants and to those from outside. He himself had found that people who knew nothing about Nola, on hearing its name and having no sense of what it was actually like, tended to indulge in a series of inventive fantasies about the place.[3] What Leone proposed to do in his work, therefore, was not simply to provide a fully documented *descriptio* of the physical appearance of the present-day city but to restore, for its inhabitants, an image of the place in all its historical and cultural complexity, enabling them to establish relations with outsiders with greater confidence, now aware both of their own *dignitas* and that of their city as a whole. Outsiders, for their part, would be more familiar with the prestigious history of the place and would treat its inhabitants with greater respect. It was a common belief that in any city worthy of note the inhabitants would show a corresponding cultivation.

---

1 Leone 1514, bk. III, ch. 3, f. xxxxiiii *verso*: "Quod summopere nunc nos cupientes conferre cum praeteritis praesentia consequi non possumus; nequimus enim comparationem facere praesentium cum anteactis saeculis; de familiis enim ut de plerisque aliis rebus Nolanis nihil a quoque relatum usque legimus". Leone's purpose in describing the Nolan families is expressed as follows: "ut illorum quisque possit iudicare quo pacto consentiant posteriora cum praesentibus, cernereque unde nam genus trahat atque an recaeptitius sit an indigena civis, et an a serviciis ad honestos evectus an a claris ad servicia degenerarit"(ibid.).

2 Leone 1514, bk. III, ch. 3, f. xxxxix *recto*.

3 Leone 1514, *Praefatio*, f. ii *verso* (see below, Appendix 2, § 7).: "audientes in quadam inani fictaque cogitatione tamquam insomnio vagantur".

Leone's work also had a specific didactic purpose: in learning of their glorious past, present-day and future Nolans would be encouraged to emulate it. The awareness of the city's ancient renown would, given the profound genetic link binding individuals to their native territories, their families and their cities, be a powerful force in dissuading them from acting badly and encouraging them to behave nobly.[4] This assumption that there is a close relation between a city's population and the biological and anthropological context in which that population is rooted is one of the concepts underlying the methodological framework of Leone's description of Nola.

It is therefore not by chance that in the dedicatory letter to Enrico Orsini Leone explicitly states that his desire to write the history of Nola is connected to his awareness that he himself is a product of the city and its civic culture.[5] The account he provides of the various facets of the city's life is developed within an ideological framework which incorporates civic pride and classical erudition and is not without an element of rhetorical display, in which the connection *respublica-cives* is brought to the fore.[6] Leone was not alone in this: in the second half of the fifteenth century a similar devotion towards their native cities can be found in other men of learning who had left the Kingdom to study at universities and undertake their professional activities elsewhere.[7] A similar sense of civic pride and of *amor patriae* can be found, for example, in the work of a contemporary of Leone's, the doctor, philosopher and astrologer Angelo Catone from Benevento (d. 1495), who graduated in Padua and went on to teach at Perugia, Bologna and Naples during the reign of Ferrante I.[8] His 1464 treatise *De epidemia*, written for the consuls in Benevento in order to inform them of the causes of plague and advise them on how to deal with

---

4 Leone 1514, *Praefatio*, f. ii *verso*–iii *recto*. See below, Appendix 2, § 13.

5 Vitale 2015, 260–261.

6 Martini 1981, 341–367, remarks that the picture of Nolan life evoked by Leone, "fired as it is by his sacred enthusiasm for the beauty of the city, the nobility of its inhabitants, the grandeur of its traditions ... borders on the celebratory and the idyllic"; although he also goes on to note that the information which Leone provides "is generally confirmed by the other sources we have" and that "the portrait of a vital and commercially flourishing city corresponds in large part to a real period of prosperity in the history of Nola".

7 For biographical information on Leone see the Introduction to this volume, 1–3.

8 On Angelo Catone, see Figliuolo 1997, 277–413. Catone led a politically and culturally intense existence: he followed the young Federico of Aragon to Burgundy (1474–1476), and afterwards went to the court of Louis XI in France. He was appointed Archbishop of Vienne (1480) and undertook diplomatic missions to Milan and Rome.

an outbreak in the city, has an introduction full of expressions of patriotic involvement and concern.[9]

But it should also be remembered that, at the same time as such treatises were being produced, another, opposing, view emerged in the humanist circles centred on the court in Naples, where the expectations and ambitions of the administrative elite were at play, which asserted that to serve the king was not merely a means of social advancement but also the ethical and political destiny of the nobility.[10]

1   Book III of the *De Nola* as a Source for Socio-political and
         Economic History

Leaving aside the objective facts contained in Book III of the *De Nola* on the administrative system, funeral rites and customs and on all other matters relating to the life of the city which could be described without the need for methodological premises or philosophical interpretation, this part of Leone's work is without doubt a markedly ideological text, especially its third chapter, *De familiis praesentis Urbis egregiis* (though this section too is a useful source of information on various aspects of the social and economic life of contemporary Nola). In this chapter Leone draws the portrait of a group of families, about 115 in all, whom he considers to be *egregiae*, the word used to describe them in the title. He claims to have singled out these families on the basis of the criterion according to which there were two social categories in contemporary Nola: the category of those who enjoyed no status of any kind and that comprised by people who stood out in local society.[11] The sources of income of this second category are briefly outlined: according to Leone's findings,

9   The *De epidemia* is conserved in Rome, Biblioteca Angelica, ms. *Angel. Lat.* 1371, ff. 1 *recto*–69 *recto*. An Italian translation of this work is provided in Catone (ed. Baldi et al.) 1968.

10   See Tristano Caracciolo's eloquent reflections on this theme in *Plura bene vivendi praecepta ad filium* (1509 ca.) where he remarks "si sint qui patriam praeferendam putent tamquam totum parti, mihi autem patriae et reipublicae nomen in principem iam cessisse videtur; porro quid quid sit et patriae nomen obtineat pro virili parte iuva et orna"; and in the *Neapolitanae nobilitatis defensio* (1480 ca.): "Magnos certe quaestus ex principum obsequiis, quae nostra mercatura est, percipere solebamus". See Vitale 2003, 166, and Vitale 2002, 102; see also Delle Donne 2007.

11   According to Martini's calculations, Leone's choice of families represents about 20% of the number of families resident in Nola, a fairly high proportion, in other words, and one which shows the widespread level of prosperity and cultivation in Nolan society. See Martini 1981, 341–367. See Spampanato 1921, I, 10 for Nola's demography in the mid-sixteenth century.

the economic status of this social category was founded, in addition to their earnings from the various professions they exercised[12] and from the produce of their agricultural estates, also from the sums of money, quotas of a feudal nature, which were granted by the sovereign on tax payments made on the use of buildings and land, in line with a widespread practice, also found in the capital, which Leone takes it upon himself to explain to his readers.[13] It seems that the income from such concessions was not especially high, never amounting to more than 300 ducats annually for each individual. Leone on the contrary devotes much attention to the professions practised by each family, seeing them as the principal ways in which these families acquired their standing in the city.[14]

On this point it is interesting to compare Leone's outlook with the view of the nobility expressed by Francesco Elio Marchese (d. 1517), writing in about the same period—the second half of the fifteenth century—as Leone, in his account of the *anoblissement* of numerous Neapolitan families. Given the difference in the two authors' political and cultural viewpoints, the image they draw in their writings of these two aristocratic castes is profoundly different. Leone's focus is on the contemporary city and he therefore devotes no pages to analysing the problematic ways in which the aristocratic class he is describing developed and evolved over time; he does not examine the influence of single individuals or of family groups on the way political power was managed, merely limiting himself to listing the ecclesiastical or administrative positions, both locally and in the capital, which several Nolans occupied. Nor does he take into consideration the role played in these families' socio-political ascendancy by such factors as their exercise of feudal privileges over individuals and possessions or their strategies of dynastic intermarriage, both fundamental elements in the political and economic mechanisms of *ancien régime* societies.[15]

---

12   In the listing of the professions and trades exercised by the citizens of Nola found in Ruggiero 1997, 54–64, the majority (71) are military, followed by men of letters (52), merchants (47), jurisconsults (46), clergy (24), musicians (8), orators (7), monks and preachers (7), doctors (4) and surgeons (3). In addition there are 10 philosophers, 4 grammarians, 2 poets, 4 horse-breeders, 1 architect, and 4 surveyors.

13   Leone 1514, bk. III, ch. 8, f. liii *verso*: "Pecunia autem privata civibus exigitur, partim a feudis, partim ab agrorum fructibus; feudum enim est ius quoddam regium, quod habetur in plerisque aut aedibus aut tabernis aut agris, aut omnibus his, quorum feudatarii tenentur quotannis tributula quaedam pendere feudi domino". See Vitale 2003, 37–48.

14   A bibliography on Marchese can be found in Bianca 2007.

15   In connection with the socio-economic role of marriage, Leone mentions the law, no longer in force in his day, which had been issued by Raimondo Orsini, Count of Nola († 1459), which prohibited the women of Nola marrying men in other cities and therefore taking

The dominant social category in Nola as described by Leone cannot by any stretch of the imagination be designated as "feudal".

Given this, it should be unnecessary to underline that, from a methodological point of view, Leone's description of Nolan nobility cannot be taken as either the only—or even the most comprehensive—source for an overall analysis of the socio-economic structures in general of the city and its ruling class without the corroboration of other and different kinds of sources. The information provided by Leone is generic and it is hard to assess its degree of subjectivity. The valuation is made much harder by the limited availability or even, at least at present, the complete lack of documentary sources (as Giuseppe Martini pointed out as long ago as 1981) on even fundamental elements in Nolan society: on the degree of its involvement in agricultural development and its management, on the extent of the commercial networks which the city encouraged, and on the individuals who were engaged in such activities and in other sectors of the economy, and on their participation in the political and administrative activities of the *universitas* (the principal association for citizens) in Nola, both as far as relations with the Orsini family, the feudal rulers of the city since the late Middle Ages, and relations with the institutions of central government and the monarchy were concerned.[16] Nor do private archives survive in which it might be possible to trace the rise of certain families belonging to the city's elite or the general context in which they emerged into prominence.[17]

---

their dowries away from Nola, whereas the opposite was allowed. The law was inspired by the desire to protect and manage the wealth of local families with prudence. See Leone 1514, bk. III, ch. 9, f. liiii *recto*.

16   The documentation relating to this period in the series of texts of the *Consiglio Collaterale* and the *Partium* of the *Sommaria* suffered serious losses during the Second World War as the result of a reprisal by Nazi troops. The series of *Conti erariali*, in the *Dipendenze della Sommaria*, from which according to Martini it would be possible to extract information on which individuals occupied the roles of *erari* from 1481 onwards, were unavailable owing to internal reorganisation of the State Archives in Naples. See Martini 1981.

17   Information can be found in the holdings of the Archivio Diocesano in Nola studied by Buonaguro 1997 and from those documents published in the series of *I Registri della Cancelleria angioina* reconstructed by Neapolitan archivists under the direction of Riccardo Filangieri, as well as from the manuscripts in the Biblioteca Nazionale di Napoli (especially Carlo de Lellis' unpublished notes on Neapolitan families held in mss. X.A.1, X.A.2, and X.A.3). Other useful sources are the manuscript notes to catalogues of the archival material relating to the Angevin period in the State Archives of Naples (see, for instance, Sigismondo Sicola's *Repertori*), and the family and local histories compiled by antiquarians in the sixteenth and seventeenth centuries, the information in which needs

2 Social Topography and Types of Residential Dwelling

It is far from easy to identify the typology of family dwellings within the urban space of Nola in all their socio-economic variety from the pages of Leone's account. It is probable that their organisation corresponded to the socio-anthropological patterns of residences found in Naples and other centres in the Kingdom but without the codified values found there, at least as far as the close connection between a family's place of residence (and how long they had lived there) and their position within a political or administrative institution such as the 'Seggi' (positions which brought with them privileges and prestige) are concerned.[18] Leone does not refer to the existence of any rules governing such connections and nor can further details be found in the surviving documentation.

The selection and election of members of the city's administration involved the whole of the populace wherever they lived within its walls but on the basis, as already pointed out, on whether individuals belonged to one or other of the two social categories of *primarii* or *plebeii*. This view is confirmed by the existence of a single building where the citizenry could assemble. Leone adheres to his topographical analysis of the city by noting the presence of a *Porticus* without however explaining what it was used for, which he takes for granted, although he gives a detailed description of its position and its style of architecture [fig. 33].[19] This is similar to the many other 'Seggi' familiar to us in other cities in the Kingdom: an edifice standing a few steps above street-level, with a stone bench running the length of the perimeter, where people could hold meetings.[20] The importance of the *Porticus*, with its central position opposite

---

of course to be treated with caution. However, the documentary information contained in the eighteenth-century work of the Nolan prelate Gian Stefano Remondini 1747–1757, and in the work of Vincenti 1898, remains valuable. Vincenti's work, as well as containing an account of the Orsini dynasty, opens with observations on the city's economy, administration, and social life.

18   On this subject see Vitale 2003, 147–153.

19   Leone 1514, bk. II, ch. 10. On the function of the *portici* (i.e. the seggio buildings) in Naples the views of Camillo Tutini (1644, 37–38) are well-known; they echo those of Pontano's *De magnificentia* (cap. II) and Fabio Giordano's *Historia Neapolitana* (Biblioteca Nazionale di Napoli, ms. XII.B.26, transcribed in Schipa 1906 and in Rea 2011–2012). On the activity of the Seggi in Naples see Vitale 2010, 71–95. An accurate analysis of the structure and functions of the Seggi, drawing on a rich variety of sources, both iconographic and archival, and part of an overview of the Regno can be found in Lenzo 2014.

20   Leone 1514, bk. II, ch. 10, f. xxx *recto*: "In media urbe tethredra quaedam extat, quod Segium vocatur a sedendo. Mihi vero non tethredram hexedramve, sed porticum liceat appellare".

the cathedral, is evident also from the map of the city and its function can be likened to a town hall designed for the meetings of members of the Council which governed the *Universitas*, in the case of Nola a council consisting of six "elected" citizens, three chosen from among the *primarii* and three from the *plebei*. The council, whose members had a four-monthly mandate, was responsible for the public administration of the city and was its highest authority.[21]

Arranged according to the clan system (Leone's use of the term *tribus* to define the neighbourhoods in Nola reveals his awareness of this form of family aggregation), only in a few cases can offshoots of the same family be found detached from the main family nucleus and living in another street.[22] The connection between urban space and social organization is shown by the fact that various streets are named after the important families who lived there. For example, Via Scriniaria certainly derives from the family of the same name who lived there, who are documented in Naples in the second half of the thirteenth century and were in Nola in the following century.[23] Similarly the denomination of the street called Chiarastella is clearly connected with the family of the same name, who were active in Nola as early as the fourteenth century and whose position, as documents show, was prestigious long before Leone lists them among the *egregiae* of the city.[24]

For the many synonymous of the word 'seggio' (*porticum, theatrum, sedile, tocco* etc.), see Lenzo 2014, 9–17; 43–45.

21   Leone 1514, bk. III, ch. 7, f. li *verso*. It must be pointed out that Leone never calls the citizens who occupied the highest social positions in Nola *nobiles*, however they attained their position, but *primarii*. The information Leone gives on the date of construction of the *Portico*—he writes that Count Niccolò Orsini († 1399) was responsible—is confirmed in a document from 1390, in which reference is made to a house which stood near the *theatrum* (i.e. the seggio) and the town square; see Buonaguro 1997, 125, n. 382.

22   Such behaviour can be seen in such families as the Tansillo, the Scrinario, the Mastrillo, and the Grifo who have a residence both in the *regio* Cortefellana and in Vicanziana, as well as the Mennato in the Samuelitana and the Vicanziana. Leone 1514, bk. III, ch. 3.

23   The Scrinario (Leone 1514, bk. III, ch. 3, f. xxxvi *verso*) were an established part of the administrative nobility of Naples and had been taxed on the proceeds of their feudal estates from the mid thirteenth century onwards. See Vitale 2003, 38, 48, 81 and tables III and V. Numerous members of the family are present from the beginning of the fourteenth century as landowners in local documents. See Buonaguro 1997, ns. 102, 111, 133, 177, 272. On the Tansillo family: Leone 1514, bk. III, ch. 3, f. xxxvii *verso* and Ruggiero 1997, 447 note 11. See in particular on the church in Nola the observations of Giovanni Vitolo's introduction in Buonaguro 1997, XV–XVI.

24   Buonaguro 1997, nn. 192, 197, 418; for references to the name Chiarastella as early as the fourteenth century, see n. 284, 334, 360, 381, 390, 394, 398.

However, Leone makes no reference to any social differentiation between the various neighbourhoods and the various streets. Only in the case of the Vicanziana does he distinguish it as *nobilior*, but not because aristocratic families lived there, but for material and objective reasons, since it was the most important thoroughfare in the city.[25] As can be seen on the plan of Nola [fig. 4], the street is a fundamental part of the urban infrastructure, connecting Nola with its territory and running through the city from the Porta Samuelina to the Porta Vicanzia, next to an *arx* with a tower, in the direction of Naples. All inhabitants and visitors wishing to cross the city had to use the Vicanziana. Along its route, into which other important streets converged, such as del Portello, Chiarastella and Cortefella, could be found the customs house, the cathedral and the *episcopium*, the main square, the church of S. Felice and the Portico. That the city's leading families chose to build their residences there is a reflection of the street's importance in urban life.

In this connection, it should also be pointed out that Leone attributes no particular importance to the *regio* Portellana, even though the Orsini had their palace there, the "regia" [fig. 28]. The Orsini are listed as one of the seven families with residences in this neighbourhood, two of which were renowned: the Correale, whose rise to prominence in the fifteenth century took place under the Aragonese dynasty, and the Albertini, perhaps one of the most important families from the early sixteenth century onwards on account of the positions they held in the central government of the Kingdom and the extent of their estates.[26] The question arises whether Leone gives little emphasis to this area of the city because it was the smallest in extent or, perhaps above all, because it played no practical role in the city's commercial life: the gate in the stretch of city walls which bounded the neighborhood was closed (the *Porta clausa*) and no transit was possible.

The type of residential house described by Leone in chapter xv of Book II was certainly based on the model of the *domus*, corresponding to a contemporary form of an elegant urban residence occupied by the elite in the areas of greatest social prestige. It indicates a comfortable tenor of life, a culturally refined way of familial and social living, as well as the desire for an aesthetically pleasing organization of daily existence, as Leone's description of the functions of some of the rooms allows us to glimpse: the upper loggia was for conversation and a place from which one could enjoy the view and the cooling breezes; in the library members of the family could go to study, meditate, and

---

25   Leone 1514, bk. II, ch. 10, f. xxx *recto–verso*.

26   On the Correale (or Curiale), who were favoured by Alfonso of Aragon, and held leading
      positions at court under the Aragonese, see Aldimari 1691, III, 622.

discuss.[27] This image of the city described in considerable realistic detail, is one of the many signs of the contemporary, almost documentary-like vividness which Leone gives to his text.

3 A Society Open to Social Mobility

An examination of the list of families drawn up in the *De Nola*—which is perhaps not an exhaustive one, comprising the entire local aristocracy, since, as mentioned above, Leone makes a selection based on his personal conception or model of nobility—reveals that numerous families had their origins in other cities. A significant number came from Amalfi and Sorrento, but there were also families, probably branches which had separated from the main nucleus, from Naples and Salerno (Palma, Sasso, Frezza, Bonito, Correale, de Gennaro, del Giudice, de Riso, Caracciolo) or from Italian cities outside the Kingdom (Mazzei, Albertini, from whom the princes of Cimitile were descended, or Mastrillo).[28] In drawing attention to the presence of Genoese and Venetian merchants in Nola, Leone emphasizes their lively commerce with their respective homelands and the great wealth they accumulated.[29]

Such territorial mobility among Nola's citizens reflects a wider phenomenon in the cities throughout the Kingdom as recent studies reconstructing the formation of elites from the Angevin-Aragonese period onwards have shown.[30] But the factors at play in a family's rise to prominence were very diverse. In differing, complex and continually evolving socio-economic and political contexts—sometimes long-lasting, sometimes due to a particular combination of circumstances—individuals and families from various geographical

---

27   Leone 1514, bk. II, ch. 15, f. xxxix *recto*. See de Divitiis' in this volume, 92–100; Appendix 3, 166–168.

28   On the Palma family see De Lellis (ed. Conforto) 1701, 71–93. On the Sasso family, see Aldimari 1691, III, 717–720. Aldimari writes that in 1496 Notarino was *maestro razionale* in the Camera della Sommaria, while his son Mario was a member of the Regio Consiglio di S. Chiara; he states that in this period the family belonged to the Nolan nobility. On the Frezza family, see De Lellis 1654–1671, III, 161–169. The Frezza were one of the Amalfitan families who acquired importance by financing the Crown and held the offices of *secrezia e portolanato* in various cities as well as posts in the central administration. In Naples, the family obtained the right to belong to the Seggio di Nido and to build a chapel in the church of S. Domenico Maggiore: their descendants went on to occupy high offices under the Angevins.

29   Leone 1514, bk. III, ch. 5, f. l *recto*: "ii ditissimi in patriam rediere".

30   Vitale 2003, 83–124; Vitale 2009, 33–53; Vitale 2015, 273–278.

and cultural backgrounds could trigger processes which over time transformed the physiognomy of their local communities, as the effects of the diaspora from Amalfi or the influx into the Kingdom from other Italian cities and regions (Venice, Florence, Lombardy, etc.) show. Even if we look at just a limited part of the Kingdom, Puglia, we can find immigrants settling in such urban centres as Trani, Barletta, Bitonto, Brindisi, etc. and going on to play an active role in their economic, political and administrative life, and also becoming part, often a very important one, of their ruling elite.[31]

There are various indications that Nola was not only a focus for commercial exchange and a bridge between the agrarian hinterland and the coast but was also part of the circuit of Mediterranean trade, exporting to Egypt part of the city's abundant hazelnut production, and importing in exchange at huge profit provisions of spices.[32] Among his brief remarks on the mercantile spirit of the local aristocracy in Nola, Leone singles out the Chiaramonte family, whose story seems to sum up, almost emblematically, this aspect of the city's socio-economic activity. He recounts the initiative of one of its members, Pacello, in promoting an import-export business with Egypt through his son Angelo and also its involvement in importing and cultivating palm trees and pepper plants (he mentions a palm tree in the garden of the Chiaramonte house which has grown as high as a tower).[33] But other individuals are mentioned who, encouraged by the fertility of the Nolan countryside—a feature Leone singles out for praise on several occasions—which guaranteed them success, were interested in agricultural experimentation, such as Antonello from the Campobasso family who introduced the successful cultivation of saffron

---

31   It has proved possible to reconstruct the biographies and careers of numerous individu-
      als who took part in the diaspora from Amalfi from the Swabian and Angevin periods
      onwards. They were both merchants and money dealers, ready to exploit the considerable
      fortunes they amassed and their own expertise by putting these at the service of the sov-
      ereign and increasing their wealth even further. They were responsible for managing local
      and regional offices, while, in the wake of the institutional and administrative reforms
      which took place under the Swabians, they succeeded in gaining control of key sectors
      and formed a powerful group in what was an organised economic system, sometimes
      shared between families. For further details see Vitale 2010 and Vitale 2015, 82–109.

32   Leone speaks of the port of Stabia (in other words, Castellamare di Stabia) as the natural
      outlet for goods produced in Nolan territory, but it has long been argued that this is in
      fact the ancient port of Oplonti, the present-day Torre Annunziata, which was connected
      directly by road to Nola; on the question see Martini 1981, 309.

33   Leone 1514, bk. I, ch. 2, f. vii *recto*; bk. III, ch. 3, f. xxxxviii *recto*; and see also bk. III, ch. 5,
      f. l *recto*.

from the Abruzzo.[34] Leone himself planted cinnamon trees, which went on to flourish.[35]

The presence of the Orsini family, who played an important role in the politics of the Italian peninsula, gave rise to numerous opportunities for connections with other cities but also enabled various individuals to take on administrative offices at their court. For example, Leone, when he writes about the Alfano family, mentions, among others of its leading members, Giovanni, who was so shy that he abstained from any participation in the public realm. Leone remarks that he was a favourite of the count Orso.[36]

It is impossible here to describe the biographies of the numerous families mentioned by Leone, though information on some of these figures, drawn from archival sources and genealogical scholarship, can be found in the notes.

Relations between Neapolitan and Nolan society must have been very close; various families—or their branches—from the capital became part of the Nolan nobility, though they still, especially in the sixteenth century, maintained their Neapolitan residence and defended their right to belong to their local Seggio, which guaranteed them certain privileges. Social customs such as the funeral rites described by Leone were identical to Neapolitan ones: the descriptions of ritual behavior, of ceremonies, even the custom of building a 'castellana' (or catafalque) for funerals as well as the style, evoking classical models, in which it was constructed, correspond exactly to those described in chronicles of daily life in the capital city.[37] Did Nola imitate Naples or vice-versa? Or did the customs arise from a shared cultural background?

4   Leone's Cultural Model of Nobility

The historiographical approach adopted by Leone—the choice of restricting himself exclusively to a direct, almost eye-witness, knowledge of Nolan society—undoubtedly sprang from his scientific training, his interest in the

34   On Antonello Campobasso's initiative, see Leone 1514, bk. I, ch. 2, f. vii *recto*. The
       Campobasso family is mentioned in bk. III, ch. 3, f. xxxxvii *verso*.
35   Leone 1514, bk. I, ch. 2, f. vii *recto*.
36   Leone 1514, bk. III, ch. 3, f. xxxxvi *verso*. Giovanni d'Alfano's activities in the service of the
       count are documented in the manuscript of the State Archives of Naples, *Dipendenze
       della Sommaria, Conti erariali*, I Serie, Fasc. 639: "Libro facto per me Iohanni de Alfanis
       continenti tucti denari che se receperanno per mi da li erari et camerlenghi de Nola"
       [1490]. In 1457 Giovanni d'Alfano was secretary to the Count of Nola; see the manuscript
       Biblioteca Nazionale di Napoli, X.A.1, f. 176.
37   Vitale 2006, 115–119.

study of physical facts and the measurement and examination of material specimens (it should not be forgotten that he was also interested in nephrology and therefore in the analysis of urine and other experimental observations).[38] While for his account of Nola during antiquity and the paleochristian period, Leone relies, in addition to archeological evidence, on written sources, Classical and early Christian literature and epigraphy, for his description of contemporary Nola he leaves aside the study of how it emerged and calls instead only on his own experience, excluding all written sources, including diplomatic accounts.

In short, Leone gives us a snapshot of reality at a certain time and in a certain place, a crystallised image of the social structure of Nola as he knew it and which he proceeds to analyse by selecting a number of families to discuss on the basis of a model according to which pre-eminence is justified solely through the attainment of excellence in various professional activities—in medicine, letters, music, commerce and arms. Outside these professions, in the exercise of which true excellence can be achieved, and ignoring other traditional conceptions of nobility, Leone considers the Sarno and Mennato families worthy of mention for their distinction in breeding horses, as well as the Coci, who, although they are neither soldiers nor men of letters, never debase themselves with servile activities.

The conception of nobility which inspired Leone deserves detailed analysis, but, by way of conclusion, I will restrict myself to the consideration of the significant view he expresses in another of his works, the dialogue *De nobilitate rerum*, published posthumously in 1525, in particular in chapter XXXII, entitled *Disputatio cum Bruto Longo an divitiae et claritas avorum nobilitent*.[39] In discussing the concept that neither wealth nor birth, being precarious and changeable goods, can constitute a stable basis for human character and cannot therefore be regarded as defining elements of nobility of character, he states that riches are never ours either when we possess them or when we lose them, so that the man who possesses wealth is no different from what he is when he loses it.[40]

A detailed comparison of Leone's views on nobility with those expressed in the abundant contemporary literature on the topic can be found in the studies

---

38   Leone 1519.

39   Leone 1525. For an overview of the different positions which men of learning took on the question during this period of transition, see Donati 1988, 3–92.

40   Leone 1525, cap. XXXII: "Divitiae neque quando praesentes sunt, neque quando abfuerint, nostrae sunt".

on this argument.[41] As far as Leone's numerous and detailed examples taken from the world of nature in the *De nobilitate* are concerned, the way he carries out the evaluation of excellence in Nolan families is analogous to the framework of the dialogue as a whole, which takes the form of an Arcadian idyll, a conversation between Leone and some friends who, on a hot July day, in order to escape the "insanus Solis aestus" take themselves off "ad viridaria" on the outskirts of the city, strolling through the gardens and orchards of the "vicus Sancti Pauli", where their young friend Vincenzo Chiaramonte owns a vast and fertile estate and has prepared a supper for the group. After their learned discussion has concluded, the participants return "ad citrorum pergulas".[42]

Leone makes his judgements with irony, not merely on the theme of 'noble blood' but also on the reliability of written evidence. An example of this mocking tone can be seen in the disagreement between Bernardino Lepido and Bruto Longo. In denying the possibility of acquiring nobility through external possessions and advantages, Leone concludes that the nobility which derives from such external connotations is a mere shadow of nobility, thus dismissing Bruto Longo's boast, for which he claims he has documentary proof to be descended from the emperor Frederick II.[43] In Leone's story, Longo, coming from Lauro, joins the group and their discussion, spending the whole day in genealogical comparisons of family trees and ancestries.[44] The friends attack Longo's

---

41   Bracciolini (ed. Canfora) 2002; Caracciolo (ed. Paladino) 1934–1935; Landino (ed. Liaci) 1970; Galateo (ed. Tateo et alii) 1994. English translation of all these authors are given in Rabil Jr. 1991. For a general overview of the debate on nobility in the Italian Renaissance see Tateo 1974; Donati 1988; Rabil Jr. 1991; Finzi 2010.

42   Leone 1525, passim. By way of example: Leone divides elements into animate and inanimate and asserts that the former category is superior. Some of his distinctions he draws from this premise include: stones are less noble than metals, but gems are nobler. Among grasses grain is the highest element but animals are superior to the vegetable world. Animals which live on the ground are more noble than creatures which fly. The dolphin is pre-eminent among aquatic animals. In the animal world, for example, dogs, horses and lions are all examples of excellence but the highest rank is occupied by the elephant ("elephantos omnium animalium excepto homine perfectissimos atque nobilissimos esse") and among birds the eagle. Fortitude is a more perfect virtue than temperance, and justice is superior to constancy. The honest contemplative is more noble than the active. Wisdom is the highest virtue.

43   "Seque habere trahereque dicebat originem a divo Federico imperatore, idque monumentis quibusdam literarum ostendebat, quae instrumenta hodie vocant". Donati 1988, 86, in his analysis of the *Nennio* (Nenna 1542) states: "but perhaps Nenna had in mind the dialogue composed by Ambrogio Leone from Nola, in which Bruto Longo appears in a negative light" and refers explicitly to Bruto Longo's boastfulness in Leone's text.

44   Leone 1525, ch. XXXII.

point of view, saying that no man is nobler than another because of his family, so that anyone who claims to be noble because of his descent from royalty is mad. As for the nobility which derives from wealth, Leone, with the help of an image, writes that wealth and power, in their value and duration, are like the leaves on a tree which a slight wind in autumn dislodges. Wealth is stolen or lost through the envy of others, by tyrants, through shipwreck or theft or fire, because of gambling or gluttony or love of luxury or lust or idleness.

Leone's long residence outside the Kingdom (for example, in Padua and Venice) to study and to practise his profession meant he had opportunities to meet other scholars and to make the values of northern Italian urban society his own. His philosophical and scientific background, as with other men of letters who were his contemporaries, led him to take innovative approaches to the fields he studies, including urban history, approaches which also influenced the humanistic production of erudite genealogy which rejected all speculations not based on reliable documentary proof.[45]

Leone's attribution of aristocratic dignity to those who dedicated themselves "studiis ingenuis" (in the examples he gives, these are prevalently either legal studies "in utroque iure" and medical theory and practice) was, as a viewpoint, not simply limited to the restricted cultural circles in which he moved; it was an outlook which in certain ways and in certain circumstances was in line with the political system of privileges granted to social categories in the statutes of some of the *universitates* in the Kingdom, as well as with the attitudes expressed in Neapolitan treatises in the fifteenth century, in which a symbiosis of values such as *virtus* and *divitiae*, both in individuals and in families, was seen as more important for the definition of nobility than the mere fact of noble birth.[46] The statutes of a prestigious and exclusive chivalric order such

---

45   Defilippis 1991. The method Leone adopts in his description of what constitutes Nolan
      'nobility' is quite different, despite his aims to be objective, from the criteria of histori-
      cal reconstruction found in the work of various humanists who were active during the
      Aragonese period, such as Francesco Elio Marchese, whom Croce (1922) 1927 singled out
      for praise. Marchese, in his biographical accounts of noble Neapolitan families, drew on
      archival information, thereby demythologizing the past, freeing it of legendary attribu-
      tions, even at the cost of damaging the families' genealogical pride as well as their specific
      political interests. Marchese's approach provoked, as is well known, a polemical reaction
      and revision in the seventeenth century on the part of Borrelli 1653; the question was
      deeply rooted in the dialectical relation between the families and the social groupings in
      the Seggi.

46   On the subject see Vitale 2002, in particular the chapters *Nobilitas=gentilitas + virtutum
      fortunarumque copia*, and *Nobiltà e mercatura*. See also the chapter *Nobiltà e divitiae*,
      87–110; see also Vitale 2003, 155–179, chap. 4: *La costruzione dell'immagine*.

as the 'Ermellino', founded by Ferrante of Aragon in 1465, stated that that man was to be considered noble who was known for his *virtus* and his excellent way of life.[47] The Order could be conferred not only on those who had inherited their noble status by virtue of their families but also on those who acquired nobility "a se suo labore et industria parta"; it is therefore not so remarkable that what Leone valued in the families and individuals he included among the elite citizens of Nola were civic virtues, excellence in the liberal and even in the mechanical arts[48] and in musical skills (which were now a standard part of the cultural attainments expected of a nobleman).[49]

In Leone's view, the outstanding individual is someone who develops his innate capacities to the highest degree.[50] He gives a detailed analysis of this conception in the *De nobilitate* and illustrates it with numerous examples. He starts with a fundamental subdivision between animate and inanimate things, the former superior to the latter and then examines the characteristics of the various *genera* and compares their qualities in order to identify where one is superior to the others, as well as the superiority of the single subject belonging to a *genus* defined as excellent who, by actualizing all his intrinsic possibility, has attained complete perfection. Leone's use of the Aristotelian concept of the dynamic relationship between actuality and potentiality is evident. Yet it is also true that, if in the theoretical discussion of the *De nobilitate* Leone defines "nobilitas" as exclusively the expression of the *virtus* of an excellent individual, in the *De Nola* he returns, albeit with new values, to the traditional criterion of family descent, based on the naturalistic ideology of inherited nobility. In such a conception, the "lumen virtutis" possessed by an individual is not invalidated by the eventual mediocrity or obscurity of the family to which he belongs, just

---

47   Chapter XXVI of the *Ordine dell'Ermellino* stated: "Item, consyderantes ad perfectam vir-
        tutem ac gloriam et hominum opinionem multum conferre generis nobilitatem, tametsi
        satis nobilem eum esse constet qui et virtute clarus est et optimis moribus praeditus,
        decernimus hunc ordinem eis qui viri clari et nobiles fuerint, non innobilibus et minus
        claris esse conferendum, sive nobilitate a maioribus accepta sive a se suo labore et indu-
        stria parta, dummodo huiusmodi sit ut aliorum nobilium prosapiae comparanda videa-
        tur oporteatque eos, qui suscepturi erunt, equestri dignitate esse insignitos"; see Vitale
        1999, 77.
48   It now seems established fact that merchants and physicians, also in other cities in the
        Kingdom, could be integrated into the nobility. On the cases of Barletta and Amalfi see
        Vitale 2014, 129–164; Vitale 2015, and Del Treppo, Leone 1977, 121–137.
49   Vitale 2002, 40–45.
50   Beginning with the second chapter entitled *Invenitur nobilitatis diffinitio*, Leone argues
        that nobility consists in the attainment of perfection in one's own "kind".

as, vice versa, the goodness and virtue of an individual is communicated to his family and descendants.[51]

Leone's analysis, in chapter LI of the *De nobilitate* entitled *Nobilissimus hominum demonstrator*, of a range of characteristics associated with individual excellence repeats some of the arguments on this topic found elsewhere: for example, on the question of what a man required in order to be called noble, much debated in the literature, he gives prominence to an individual's intelligence and capacity to apply this intelligence in a diversity of fields, from science and architecture to the good governance of nations. Wealth and beauty, in Leone's view, only serve to make a personality shine the more, like adding a gem to a golden jewel.

---

51   Leone 1514, bk. III, ch. 3, f. xxxxiiii *recto*: "In quo nemini quoque mirum esse velim si alicuius familiae unum alterumque nominaverim virum egregium atque horum causa etiam familiam excaeperim enotarimque quamvis illa tota sordida sit. Siquidem bonitatis atque virtutis munus hoc ipsum potissimum est ut potius a viro vel modice studioso claroque familia ignobilis obscuraque possit illustrari, quam lumen virtutis viri solius valeat extingui ab obscuritate familiae, quasi minima scintilla facilius magnas tenebras quam tenebrae maximae parvam scintillam vincere atque superare posse". It would therefore be incorrect to sustain that Leone was directly opposed to the principle of dynastic transmission.

CHAPTER 7

The Elegance of the Past: Descriptions of Rituals, Ceremonies and Festivals in Nola

Eugenio Imbriani

In the last section of the third book of the *De Nola*, in chapters 11–14, Ambrogio Leone examines the customs of the inhabitants of Nola and, in particular, the ceremonies and festivals which took place in the town. In this contribution we will look at this part of Leone's work in the context of the popular folklore and culture of southern Italy. Adopting an anthropological approach, we will show how, in accordance with his praise of the town and of its inhabitants in the preceding parts of his book, he finds cases and examples in the past to which he ascribes the origins of the ritualised behaviour of present-day Nola, with the aim of dignifying these rituals even when on the surface they appear to be merely coarse or violent.

1 Disparities

With the aim of studying the phenomenon of what he has termed the differences of level or disparities (*dislivelli*) found within cultures, the Italian anthropologist Alberto Cirese has sought examples of this not only in archival sources but also in literature, philosophy and chorographical writing and has found abundant material for his theories in the Renaissance period, especially but not only in Italy. The inequalities of level of which he speaks are constituted, as far as cultural production and reception are concerned, by the various groups which have historically made up different societies. The higher social echelons enjoy the advantages of education and of knowledge acquired on scientific grounds, in addition to greater prosperity, compared with the poorer classes whose behaviour is more closely tied to custom and practical experience; these distinctions apply also in the religious sphere, although shared behaviours and beliefs are also present. In the fifteenth century, especially in the second half of the century, this situation became more apparent than it had been in the past, taking forms which both described and represented their contexts. The interest in the popular world and the turn towards it is distinct from both Arcadian idealisation and caustic satire on peasant life but neither does it imply any participation in the social life of the lower classes.[1] In Cirese's view, the views of humanist circles in Naples exemplify the limits which were placed on the benevolent attitude they assumed towards the life of ordinary people. He refers in particular to the circle round Antonio Beccadelli, called 'il Panormita' and the founder of what would become known as the Accademia Pontaniana after the name of Beccadelli's friend and disciple Gioviano Pontano. Pontano portrays Beccadelli in the dialogue entitled *Antonius* joking with common people, listening attentively, along with his friends, to some verses being recited by a strolling ballad-singer, and, on the point of death, reciting, in immaculate Latin, a curse on rabies, a widespread disease in Apulia. He pays another ironic compliment to the Apulians for inventing, in order to justify the filthiest and most shameful behaviour, the legends of the tarantula spider's bite and the therapy needed to treat it:

> But the Apulians alone had a ready excuse for their insanity: to wit, the spider they call the tarantula, whose bite drives people crazy [...] There were spiders with different poisons, and among them even some, called concubinary spiders, that stirred people to lust. Women liked to get bitten by this spider as often as possible, and then it was permissible and lawful for them to seek men freely and with impunity, because the poison could be eliminated by no other means; as a result, what was a disgrace for other women was a cure for those of Apulia.[2]

Pontano's dialogues were published from 1491 onwards and were well known in the circles associated with the Accademia; they recorded, while duly noting the differences, the shared attitudes and feelings of the members. It is interesting to trace the web of reciprocal connections, references and allusions linking

---

1 Cirese 1997.

2 Giovanni Pontano, *Dialogus qui Antonius iscribitur*, in Pontano (ed. Haig Gaisser) 2012, 127–129. Pontano (ed. Privitera) 1943, 51: "Apulos vero solos paratissimam habere insaniae excusandae rationem: aracneum illum scilicet, quam tarantulam nominant, e cuius ammorsu insaniant homines [...]. Esse autem multiplicis veneni araneos atque in iis etiam qui ad libidinem commoverent, eosque concubitarii vocari; ab hoc araneo ammorderi quam saepissime solere mulieres licereque tum illas fasque esse libere atque impune viros petere, quod id venenum alia extingui ratione nequeat, ut quod aliis flagitium, mulieribus id Apulis remedium esset". This passage was noticed by Ernesto de Martino in his book on tarantism. He finds a kernel of truth under the tone of sarcastic scorn in the interpretation attributed to Beccadelli by Pontano: it is not an illness brought on by poison but a custom—comical, superstitious, execrable, to be sure—and even an example (laughably so) of the highest form of happiness. See de Martino 2008, 200–201.

their works. In such a context, a comparison arises almost inevitably between the passage extracted from the *Antonius* and the unadorned report found, about twenty years later, in Galateo's *De situ Iapygiae*, in which the author, on returning to his native land, the Terra d'Otranto, appears to want to qualify Beccadelli's remarks:

> Nature created in this region an extremely dangerous spider, the poison from which produces effects which can be treated with the sounds of flutes and drums. I would never have given credit to such a story had I not seen it with my own eyes occur on several occasions and if I had not read in the work of Aulus Gellius, who takes it from Theophrastus, that there are certain snakes whose poisonous bite can be treated by singing and the sound of flutes.[3]

Here, along with the absence of the older humanist's mocking tone, there is no reference to either tarantulas or the licentiousness of the women; Galateo shifts our attention towards the aspect of musical therapy and is also more concerned to prove the truth of what he says and of the effects of the poison, by referring to his own experience and to the authoritative testimony of Theophrastus found in a famous chapter of the *Noctes Atticae* by Aulus Gellius. It was then common practice to cite the ancient authors as the basis for interpreting contemporary reality: our view of the present had to be founded on the glorious and noble examples, along with the thinking, which have come down to us from the ancients, whose works it was now possible to study.[4] Galateo's epistle is characterised less by irony than by the wish to find a more temperate tone, to appeal to common sense, to maintain an *aurea mediocritas* which would seek out, where possible, reasonable explanations for beliefs (she-demons, ghosts, witches, visions) commonly held not only by the ignorant but also by all those who confused reality and appearance and were unaware of the path laid down by philosophy (in other words Aristotle), that we should test phenomena with our reason. In short, ridiculing the beliefs of others does not help to understand and even less to correct them, or oppose them if they are

---

3 Galateo (ed. Defilippis) 2005, 18: "Genuit hic natura arachneum animal nocentissimum, cuius venenum fistulis, et tympanis pelli non crederem, nisi per plurima experimenta didicissem, legissemque apud Aulum Gellium, auctoritate Theophrasti, esse quosdam serpentes, quorum venenum cantu, et fistulis pellitur". Galateo's short work, as is well known, was written in about 1510 in the form of a letter and published posthumously in Galateo 1558.

4 See Imbriani 2004, especially chapter 4. Galateo said that he was a Greek who had been born in Italy by mistake.

obviously false and horrible in their consequences, such as, Galateo adds, the myth of vampirism, which has taken hold of the entire East and leads people to dig up the corpses of those suspected to be vampires and burn their hearts.[5]

Yet even with these differences the circle of intellectuals round Pontano share a certain homogeneity of attitude towards the customs and opinions found within society. Ambrogio Leone belonged to these circles, he was on familiar terms with both Beccadelli and Pontano and frequented other members. As for his relations with Galateo, Defilippis points out that the two men had "the same professional background and frequented the same political and cultural ambiences—the Aragonese court and the Accademia Pontano—and also held friendly conversations in Leone's house in Nola",[6] at which Pontano himself may have been present. It is also the case that doubts have been expressed over Leone's familiarity with this circle and the credit he enjoyed among them: "Cette Compagnie, qui accueillait dans son sein un grand nombre d'écrivains venus de toutes les Régions de la Péninsule et comptait même des membres étrangers, ne lui ouvrit jamais ses portes. J. J. Pontano, qui le présidait et qui illustra les noms de ses amis, qui en étaient les membres, à chaque page de ses oeuvres en prose et de ses poésies latines, observe à l'endroit d'A. Leone le silence le plus absolu".[7] Yet, despite the wish to distance Leone from this ambience, he remains connected to it, as he himself tells us and as almost all his commentators acknowledge was the case.[8]

2 *Servant Nolani mores antiquos*

Leone wrote his book in 1512 in Venice where he had been living for eight years. Leaving aside his claims that the work is guided by principles of exactitude, the problems relating to the sometimes confused recovery of information about his native city in the text are in part due to the author's remoteness from his sources but also in part to a not yet fully developed historiographical approach. The author's intention to write an objective description is also accompanied by an explicit wish to praise his birthplace, to write a *laudatio*, based not only on the upright way of life of the city's present inhabitants (about 4,000 at the time

---

5 Galateo 2005, 94. There is a vast bibliography on the argument. One important work is Genesin, Rizzo (eds.) 2013.
6 Defilippis 1991, 31.
7 Montera 1934, 196.
8 It is almost certain that Leone lived in Naples for a period of time and was able to develop his contacts with intellectuals in the capital city: see Ammirati 1983.

Leone was writing) but also and above all on Nola's Greco-Roman origins.[9]
The ancient city is a point of reference for the present-day centre, from which
it derives, albeit in a reduced size and with more indeterminate boundaries,
and of which it preserves memories in the form of monuments and other testimonies. The ancient city was six times the size of Nola at the beginning of
the sixteenth century, which has the shape of a half footprint. But Nola is both
the inheritor and the custodian of its ancient fame and glories: the inhabitants
speak better Latin than is spoken elsewhere, they are honest and loyal to the
point of heroism, they cultivate the arts, their land is fertile, and God looks
favourably upon them. The first of the three books which make up Leone's
work—the one on ancient Nola—concludes almost surprisingly: the Nolans
are a beautiful populace because they love to be elegant in everything they do;
all traces of coarseness and dirt are relegated to the farms which dot the surrounding countryside, where agricultural implements are stored (and where,
we might add, the peasants and livestock live): "In the city, however, all this
has been removed and only the cultivation of elegance and urbanity remains".[10]
The separation is sharp and the author makes no effort to conceal his prefer-
ences. The condition of nobility is not simply a matter of family breeding and
dynastic succession but is the consequence of a moral adherence to the city's
long history and, moving even further back in time, to maintaining a link with
its founders. Furthermore nobility depends on the love of one's native city and
the merit acquired in seeking to honour it. It is also the case that Marino Leone,
Ambrogio's father, a "tradesman descending from tradesman",[11] could not boast
noble descent. Peasants however are excluded from this vision of Nola, clearly
too busy working on the land and tending animals to cultivate themselves: they
form, so to speak, a merely functional class, to use recent terminology.[12] Leone
confirms the primacy of the urban over the rural, though he praises the good

9   Leone 1514, bk. I, ch. 3, f. ix *recto*: "Sunt itaque Nolani tum Romani tum Athenienses, utra-
       vis gens sub sole praestantissima. Verumtamen sicuti monet Strabo, prima ac antiqua
       Nolanorum origo Graeca et Attica fuit, deinde miscella facta Graecorum Latinorumque
       hominum Romana quoque habita est": ("Therefore the inhabitants of Nola are both
       Roman and Athenian, both pre-eminent races on earth. Yet, as Strabo tells us, the earliest
       origins of the Nolan people were Greek and Attic: after the fusion of the Greeks and the
       Latins, the Nolans were also considered Roman.").
10   Leone 1514, bk. I, ch. 15, f. xxi *recto*: "Hac autem amota, reliquae solae politiae et urbani-
       tates solae in urbe ipsa manebant ac retinebantur".
11   Montera 1934, 192: "marchand descendant de marchand".
12   The working class, above all the rural or peasant classes. There has been a wide debate on
       these categorisations in Italy since the Second World War and the publication of Antonio
       Gramsci's prison notebooks. For a reconstruction of the positions taken by intellectuals

organisation of agricultural activity from antiquity onwards. In a centuries-old arrangement, the farm-workers did not live within the city-walls but in huts dotted around the fields outside, in other words where they worked.

The theme is taken up again in the third book on the manners of the present-day citizens of Nola: "The inhabitants of Nola even today conserve the antique customs and elegant style of living".[13] Today, just as in the past, urban elegance and rusticity do not go together. The city is reserved for artisans, distinguished families and the clergy, accompanied by their servants, their beasts of burden and hunting animals. Such citizens cannot abide to stay too far away from the great bell-tower of the city, such is the comfort they feel living in its shadow; they dress well, wear rich jewels, are hospitable and, in a noteworthy aside, they "avoid any rustic food" ("cibos rusticorum fugiunt"), another reason, probably, for their physical beauty. It would be interesting to know exactly what kind of "rustic food" they rejected so scornfully: Leone gives no details but we know that at the time a "noble diet" consisted, by definition, of meat, especially game and especially cooked with spices, while on the other hand peasants and shepherds ate cheese, which was still unappreciated in more refined forms of dining. The rustic diet also consisted of soups made from greens, other vegetables (think of the comic figure of the peasant Bertoldo and his turnips) and legumes, while the bread eaten by peasants was a mixture of grains and seed. From the pigs kept by peasants, they were able to keep, if they were lucky, the blood and the less sought-after parts.[14]

As to how the peasants of Nola behaved, Leone gives us an example at the end of book III, in chapter XIV, when he provides a colourful description of the grape harvest. The work involved was not especially burdensome and the grape-pickers found the time to exchange banter and innuendoes with each other, sparing no-one, least of all their masters; indeed, "The grape-pickers on the days they're working in someone's vineyard or indeed right through the season of the grape harvest are possessed by Bacchus, quite out of control".[15] They seemed to have lost control of themselves, eating to excess, laughing and bellowing; everything they did was out of proportion, they lifted and carried

---

and a selection of the fundamental texts see especially Clemente, Meoni, Squillacciotti 1976; Rauty (ed.) 2015.

13   Leone 1514, bk. III, ch. 6, f. l *verso*: "Servant Nolani hac etiam tempestate mores illos antiquos atque vetustas vivendi elegantiam. In urbe enim nulla rusticitas admissa est".

14   See especially Montanari 2008; Camporesi 1985; Pastoreau 2009; Finzi 2014.

15   Leone 1514, bk. III, ch. 14, f. lviii *recto*: "At vero vindemiatores ea die, qua pro quoquam vindemiam faciunt atque per totum vindemiae tempus Baccho deo pleni esse ac furere prorsus videntur".

the ladders as if they weighed nothing, they were bursting with energy or behaved like madmen: "nam omnino furentium est". They blew raspberries or shouted insults at any passer-by, magistrates, monks, no matter whom, no respect was shown. They uttered the most obscene and scurrilous remarks but once the harvest was over the obscenities stopped too and they all returned to their normal moderate and respectful selves. All the elements of a festival are present in the annual grape-harvest and the atmosphere is precisely that of a holiday festival, in particular of popular festivals in the Renaissance, when ordinary daily behaviour was not suspended but as it were heightened and daily speech became full of references to bodily functions, digestion and sex. Games and violent behaviour—not just in speech—were a fundamental and characteristic part of such festivities in Nola and elsewhere and we shall see further examples.[16] But Leone has his own interpretation of such behaviour, a rather strained one, it is true, but perfectly in line with his vision of Nola as a city indissolubly tied to its past and its remote origins. According to Leone, the peasants' behaviour during the grape-harvest has its roots in Bacchic orgies and festivals held every three years in honour of the god. In other words, there was a significant sacred aspect to these activities. The freedom allowed to the peasants in thus honouring Bacchus, moreover, shows the great respect in which agricultural labour and the efforts invested in it were held. In this way Leone bestows a positive meaning on the licentiousness of the grape-harvest, despite its vulgarity, justifying it by evoking pagan rituals with all the prestige of Greco-Roman antiquity.

This scheme of Leone's recurs. On 25th April each year, he relates, the priests of the diocese go to Nola in procession to pay tribute to the bishop.[17] They process in line, each wearing a floral crown and holding in their hands bunches of roses, accompanied by numerous musicians as they sing hymns to God and the saints. The bishop waits for them in the cathedral seated on his throne under a vaulted structure supported on columns and decorated with leaves and figures. Each priest approaches him, kneels down and offers him the crown and the roses, which the bishop will later distribute among the noblewomen of the city. People crowd around and watch the ritual as a kind of triumph ("ad hunc veluti triumphum"); the ceremony also involves the special presentation of a large white lamb with gilded horns, a gift from the clergy in Stabia. This act of collective submission, performed with choruses, flowers and displays of gallantry, in a kind of ritual celebration of Spring, also includes therefore the

---

16   See, among a vast bibliography on the subject, Burke 1972; Burke 2009; Zemon Davis 1986; Bakhtin 1968; Heers 1983.

17   Leone 1514, bk. III, ch. 11, ff. lv *verso*–lvi *recto*.

payment of a tribute in the form of a lamb doubtless destined for the bishop's kitchens. It would be possible to identify the symbolic aspects embodied in the lamb, the image of Christ destined for sacrifice or perhaps more appropriately in this case, the Ram as the zodiac sign of Aries which marks the approach of spring. But Leone prefers to look to the past to find there a meaning for the ceremony: he recalls the way Roman praetors administered justice and demonstrated the authority of Nola over its neighbours. In addition, Apollonius of Rhodes recounts the sacrifice to Jove of a lamb with gilded horns, a ritual which perhaps the inhabitants of Stabia also used to perform: thus the present-day ceremony can be seen as the re-enactment of the former ritual. Leone's reasoning is forced but, as we have seen, his laborious interpretations were focussed on finding a Greek or Roman source for what he saw. As for the roses and other flowers: did not Venus present herself for the judgement of Paris holding a bouquet of roses? This comparison seems decidedly far-fetched to us but this is the method Leone adopts.

3   Games

I have listed briefly above the characteristic features of festivities in the Renaissance period (and, it might be added, for a good while after). These are found all over Europe in the festivities which marked important dates in the calendar. The festivities in May, in particular, were celebrated with theatrical performances, trips into the woods and the countryside to gather flowers and plants, playing practical jokes of various kinds, including violent ones, sexual licence, and excessive eating and drinking. With variations, such activities were typical of all feast days, whether Christmas, Carnival, Corpus Domini and so forth; the clergy, far from being excluded, often became the leading participants, given that it was their churches and monasteries which became the settings for satirical and burlesque performances (such as the 'Festa dei folli' or festival of madmen, the celebration of the Donkey's Mass or the Drinkers' Mass) and games involving animals, often at the animals' expense.[18]

---

18   "The festival of madmen consisted in the ritual profanation of sacred places with monster or animal masks, exchanging clerical for secular clothing, dancing, dice-playing, the singing of obscene songs, profane parodies of the liturgy and the offices of the highest ranking clergy [...] In the public highways participants in the procession rode backwards on every kind of quadruped or stood on carts heaped with rubbish which they threw at the crowds of bystanders" (Falassi 1988, 10). Leone describes such practices or ones very similar to

Leone's account of the annual tribute to the bishop is decidedly anodyne, with the priests in procession like heavenly angels, but elsewhere he gives a much more raw and realistic account of what took place each year at Pentecost in the diocesan basilica.[19] On this occasion, the priests would adorn the church with flowers and spray perfumed water on all those present;[20] in the meantime, a well-fed pig, alive and greased with tallow fat, was hung by its hind legs on a rope from the ceiling. The rope passed through a pulley and was tightened or slackened by a group of men positioned above the vault of one of the chapels, so either lifting the animal up towards the roof or lowering it to the floor. On a roughly built platform a large quantity of buckets of water, balls of tow, and bundles of nettles were arranged; a group of five or six young men (*rustici* is the term Leone uses) standing at ground level try to seize hold of the pig. A battle to take possession of the creature ensues. Those below try to grab it by the head, its front legs, its ears, but it keeps greasily slipping out of their grasp, while the men above pull the rope to lift it up. Other participants throw the water, nettles and, worse, the balls of tow now on fire down on the young men trying to get hold of the pig: "Robust youths and peasants, with bare arms and legs and thighs, and wearing cushions on their heads, try to seize the animal".[21] The *rustici* have taken the precaution of protecting their heads, but the wretched animal is entirely defenceless and has to submit to the attacks to the general amusement of the bystanders who shout their encouragement. After a violent contest, the animal is caught, either because the rope breaks or one or the other side grows tired.

The existence of such games involving animals is abundantly recorded; it is enough to think of circus animals, dancing bears transported round fairs, until even fairly recent times, organ-grinders' monkeys and conjurors' doves, racing

---

them; as we shall shortly see, they were still tolerated at the beginning of the sixteenth century.

19   *Pasca rosarum*, in Leone's expression; according to a custom still found today, the liturgy for Pentecost includes the scattering of rose petals, seen as similar to the Holy Ghost's small tongues of flame which appeared above the heads of the apostles. Leone might also be referring to the feast of Corpus Christi, a week after Pentecost, in which floral decorations were very important.

20   Clearly not holy water. There are contemporary accounts of the joke of wetting unsuspecting people with the liquid contents (not necessarily water) of a bladder or empty eggshells or even buckets, though this is probably—it is hard to be certain—the case here in Leone's account. See Burke 2009.

21   Leone 1514, bk. III, ch. 12, f. lvi *recto–verso*: "Iccirco ad eam capiendam accinguntur iuvenes validissimi rusticique nudi brachiis coxis et cruribus, quorum capita cervicalibus conteguntur".

dogs and horses (the contests, jousts and quintains, of which Leone gives some Nolan examples),[22] hunting: Capturing, torturing, killing animals, as part of a contest, either to get hold of them to eat them (chickens, turkeys, pigs) or just for entertainment (dogs, cats) were typical activities on feast days and holidays, especially, but not exclusively, at Carnival time.

As Leone points out, Giovanni Pontano in his dialogue *Caronte* mentions a similar ritual game with a pig which took place in the main basilica in Naples but he describes it in a scandalised tone, considering it to be a shameful spectacle:

> in the month of May priests go in procession through the city wearing garlands like young lovers. But this is nothing. [...] When the whole population has assembled in the church, a female piglet is let down from the beams of the rooftop, bound with a rope and smeared with soap. The country folk are there, called to the sport. Then a great contest ensues, as the yokels strive to grasp the pig while those holding her suspended artfully evade the hands of the yokels and now pull the rope away, now let it loose in different directions. While this is going on, with the crowd intent on the game and cheering now one side, now the other, then, like a heavy rain, a great quantity of water, and a greater one of sewage and urine is poured from the roof in several places at the same time. Even human excrement is involved, and it doesn't stop until the peasants have taken control of the pig by force.[23]

From this we learn that it was not only water which was thrown on the competing peasants but also urine and excrement: this, in Naples, in an important and crowded church. In the *Caronte* Pontano takes up a clear and firmly maintained position against superstition (an opposition found in other of his dialogues, but expressed with less irony and scorn); he condemns it fiercely and

---

22   Leone 1514, bk. III, ch. 7, ff. li *verso*–liii *recto*.

23   Pontano (ed. Haig Gaisser) 2012, 66–67. Pontano (ed. Privitera) 1943, 25–26: "... Maio mense sacerdotes per urbem coronati incedunt, quasi amantes adolescentuli. Sed hoc quidem levius fuerit [...]. Ubi omnis populus in templo convenit, de trabibus summi tecti resti deligata porcella demittitur ac multo sapone circunlita. Adsunt agrestes ad ludum vocati. Ibi oritur magna contentio, agrestibus ut ea potiantur annitentibus, qui vero appensam illam tenent agrestium manus, manus arte lutendibus ac nunc subtrahentibus funem nunc in diversa laxantibus. Dum haec geruntur, turba ludo intenta et nunc his nunc illis plaudente, ibi quasi himber magna vis aquarum, maior iuris atque urinae e tecto compluribus simul locis diffunditur; agitur etiam humanis excrementis, nec prius cessatur quam agrestes porcella vi potiti sunt".

judges the kind of activity just described in the harshest of terms: "And those who sport with the piglet clearly show that, just like pigs in the mud, the whole human race wallows in the slime and filth of superstition".[24]

Yet, just as Galateo asserted a qualified view of tarantism, so Leone takes issue with his master's opinion. He declares that he is not in agreement with Pontano: the sordid and no doubt noisy rolling about in the dirt trying to get hold of a greased pig is transformed in Leone's account into a representation of the epic struggle of the Giants against Jove. Thus even this ritual game with the pig is ennobled if a similarity or analogy, however tenuous, can be found with an event, whether true or false, related by classical authors. Just as the Giants were struck by rain and lightning, so the pig's attackers are the targets of buckets of water, nettles and bundles of straw ablaze. The animal finally captured is all the booty that remains for the Giants who wanted to conquer Olympus instead. Leone is never afraid to build castles in the air in the search for some kind of justification which will cast an honourable light on the behaviour of his fellow citizens; an ancient ceremony, which took place in the temple of Jove, on the very same site as the episcopal basilica, in memory of Jove's victory over the Giants, is therefore the source, accepted and adapted by Christianity, for this ritual performance:

> The later Christians, who, it seems, preferred to preserve, with correction, ancient customs and make everything serve, as far as possible, a good and holy purpose, did not alter this spectacle but kept it as one which was most fitting for religion and known and enjoyed by the populace.[25]

During the Renaissance, especially in Italy, there were many festive days: in addition to the liturgical feast days in the Christian calendar, there were locally recurrent events (for fairs and markets, and the feast days of local patron saints) as well as special occasions such an anniversaries or the visit of an important personage. It should also be remembered that they could last a long time, sometimes for weeks or even months. While the end of Carnival, the pre-eminent festivity, could be fixed with greater or less precision—with

---

24   Pontano (ed. Haig Gaisser) 2012, 66–67. Pontano (ed. Privitera) 1943, 26: "Et hi qui porcellam lusitant palam faciunt aeque ac sues luto humanum omne genus superstitionis coeno sordibus volutari".

25   Leone 1514, bk. III, ch. 12, f. lvii *recto*: "Christiana vero posteritas, cui semper placuisse videtur corrigendo servare vetusta omniaque ad usum bonum sanctumque quantum licuit transferre minus id spetaculum turbauisse, sed potius ut convenientissimum religioni ac populo consuetum et gratum reliquisse".

the arrival of Lent—it was less clear when it began, since, in actual fact, it was often continuous with the Christmas celebrations.

We find a situation of this kind in Nola: from the first day of January to the first day of March, "until the days come when meat is not eaten, known as Lent",[26] youths—not all of them and more plebeian than noble—from the neighbourhoods of the three city gates (Vicanzio, Cortefella, Samuele) would form into three groups according to where they resided and fight each other, often violently, in the afternoons. They joined battle in the areas between the three gates, both inside and outside the city walls. They threw clods of earth and stones at each other and when they got closer resorted to kicking and punching: those who were unfortunate enough to succumb would be dragged mercilessly by the hair or the neck and thrown off raised embankments. The defeated would take refuge behind their local gate and on occasion were pursued; when this occurred they would launch an offensive with the help of adults who also frequently became embroiled. During the rest of the day and night, everyone got on tranquilly with their business, without any rancour, but when the afternoon came the fighting would start again. Once again, for Leone, this custom has a classical precedent: the training which young boys in classical times would undergo in order to prepare them for war. Whatever the truth of the matter, it is easy to observe that competitions and contests, games imitating war and conflict, duelling both false and genuine are frequently seen as essential components of festive behaviour. In more specific terms, the connection between Carnival and violence in the sixteenth century was a proven fact, as the literature on the subject shows. In Peter Burke's explanation, such aggressive behaviour was often ritualised in the form of football matches or fake battles or directed against weak and vulnerable targets, such as domestic or household animals or Jews, at whom, in the streets of Rome, stones and mud were thrown. More serious incidents also took place, however, and the occasion also often provided an opportunity to take revenge and settle accounts.[27] The transition itself from Carnival to Lent was seen as a struggle between the figure of Carnival, fat and red-faced, and his emaciated, penitent antagonist;[28] the epilogue to Carnival tide in many cases involved the death of the protagonist and a funeral procession, or he was put to death by being dismembered or hanged or burnt. Death and violence were prominent aspects of Carnival.

---

26 Leone 1514, bk. III, ch. 12, f. lvii *recto*: "quousque dies astinendi a viscere edendo aderint, quae Quadragesima nuncupatur".

27 Burke 2009; Bakhtin 1968. See also Baroja 1979; Gaignebet 1974.

28 Imbriani 2012.

4   The Feast of St Paulinus

Let us look briefly at the celebrations for the feast day of Saint Paulinus, which takes place on the Sunday following the 22nd of June, famous all over the world and since 2013 included in the lists of Unesco's Intangible Cultural Heritage of Humanity, as part of the Italian association 'Rete delle grandi macchine a spalla',[29] which brings together processions in Italy which involve, as in Nola, the carrying of enormous man-made structures round the streets of a town or city. These structures in Nola are known as 'Gigli' and each one is carried by a group of men known as a 'paranza',[30] the so-called 'cullatori'. The 'Gigli' are obelisks constructed of wood and variously decorated with statues and friezes made of papier-mâché; at the foot of each obelisk a band plays musical pieces, setting the rhythm for the movement of the 'paranza'. Altogether there are nine structures, eight 'Gigli' and a boat carrying an effigy of the saint; according to a legend still told today, the saint, bishop of Nola, offered himself as a hostage to Alaric and his Visigoths, still thirsty for conquest after the sack of Rome (410) as long as they left the citizens of Nola in peace, but after a year's absence he returned in triumph to the city in a boat. Each of the 'Gigli' bears the name of a trade guild: gardeners, grocers, innkeepers, bakers, butchers, cobblers, tailors.[31] This approximate and incomplete description of the procession is intended to provide a minimal idea of what is involved, since Leone in the *De Nola* gives an account of the festivities as he had often seen them. The differences between the procession in his day and in ours will easily be noted; even when it is possible to trace certain continuities over time in cultural phenomena, it is inconceivable to suppose they remain fixed and embalmed in forms which permanently stay the same.[32]

As Leone tells the story, the Vandals and the Moors, after conquering Campania, carried off many prisoners to Africa. Paulinus was taken there with a widow whose son he had offered to replace; the prince to whom he was given as a slave was impressed by his powers of prediction and let him go free with the other prisoners. The inhabitants of Nola welcomed him home with great

---

29   The "Rete" or network also includes the "Macchina di Santa Rosa" in Viterbo, the "Varia" in Palmi, and the "Faradda di li candareri" held in Sassari.

30   A *paranza* is made up of approximately one hundred and twenty men.

31   In other places near to Nola or in Naples itself, more or less at the same time of year, similar festivals involving 'Gigli' take place. On the carrying, for devotional purposes of *guglie*, *gigli*, models of temples throughout Campania and Basilicata, see Marano 1997.

32   On the festival of the 'Gigli' there is a large bibliography. See Lanternari 1988; D'Uva 2010; Ballacchino 2011; Ballacchino 2013.

celebrations and initiated the festival with processions and performances: "Up until our own times these celebrations have been held from the sixteenth to the tenth Kalends of July",[33] that is, from the 16th to the 22nd of June, slightly earlier than today.

During the fair and feast day of St Paulinus the authority of all the magistrates in the city is suspended, except for the magistrate in charge of the market, nominated by the count, who exercises his power only in this period. He is responsible for preventing or resolving the quarrels which break out among the merchants. During the days of the fair no taxes are payable, meaning that merchants crowd into the city since all transactions are exempt from tariffs. The magistrate solemnly takes his seat after riding through the streets of Nola, richly dressed and holding a sceptre, accompanied by a long procession, with trumpeters and flautists, all following the count's banner. On the eve of the saint's feast day, the procession is held again; this time with tall candles carried by the peasants and the guilds:

> The peasants come first holding scythes following like a banner a huge torch, like a column lit up and crowned with ears of grain. The torch is so large that it is too heavy for one man to carry it so it is lifted on to a litter to be borne by many men. The peasants give money for the making of the torch. Each year the part which burns down during the procession, which they call the 'cero', is not just remade but enlarged. In the same way, another torch is made from the others and each group follows its own torch in the procession. So the 'cero' belonging to the gardeners follows, crowned with onions and garlic, with the gardeners processing behind it, and then all the remaining 'ceri' of the artisans.[34]

Thus, peasants holding scythes lead the procession, following, like a banner, the extremely tall torch known as the *cereo*, lit and encircled with ears of wheat, and resting on a base carried by several men. Each year the burnt-down

---

33   Leone 1514, bk. II, ch. 13, f. xxxvi *verso*: "atque usque ad nostram tempestatem venerationes sunt factitatae idque a xvi Kalend. iulias usque x easdem".

34   Leone 1514, bk. III, ch. 7, f. lii *verso* "Primum enim agricultores cum falcibus procedunt sequentes tamquam vexillum eorum ingentem ut columnam facem accensam spicisque triticeis redimitam; haes adeo magna est ut pondus iniustum sit homini ferenti, quare in capulo quodam erecta praefertur a pluribus. Ex collata pecunia agricultorum confecta est atque quotannis augetur non solum reficitur id quod in lustrazione incenditur quod caereum vocant. Similiter fax alia conficitur ab aliis, quam in hac pompa suam quique praemittentes secuntur. Deinde ergo sequitur fax holitorum cepis aliisque corollata, quam secuntur holitores ac deinceps reliquae faces opificum".

part of the *cereo* is renewed and enlarged. There then follows the *cero* of the gardeners hung with onions and garlic bulbs and that of the artisans. Lines of monks and priests continue the procession which ends with the bishop carrying the relics of the apostles and of the martyrs, the wood of the cross, the hand of St Paulinus,[35] and accompanied by the count and the magistrate of the market. The procession is followed by the nobility and the commoners of the city, all on foot.

Games are held on the last day of the festival period. The first of these involves a race in which the participants compete naked and with their hair tied up ("nudi ac revinctis capillis"); the second is for small boys, below the age of twelve: they ride horses until at a certain point of the route they find a ring hanging in mid-air through which they have to aim a pole. At the end of the games the magistrate of the market gives up his office and normal life resumes.

It is evident that Leone failed to find classical antecedents for the *ceri* because he mentions none. They were a tribute to the saint and carrying them constituted an act of devotion; not only would it have been a risk to try to compare them to superstitious—or worse—ritual antecedents but there was no need to find a precedent in the past which brought out their positive value and, as a result, Leone does not go searching about in it for any.[36] As for the games, there is another brief reference to the classical interest in the military education of the young. It is surprising, though, that Leone makes no comment on the naked runners resembling the naked athletes of Antiquity.

In the passage from the *ceri* to the *Gigli* clearly something has been lost—the element of fire, above all. The sight, as seen from the roofs of the city, of these huge torches lighting the procession as it wound its way through the streets in the evening must have been extraordinary. The *Gigli*, now seen as part of our cultural heritage, are different, the result of a long history of transformation, of recurrent attempts to define them as cultural objects, both material and immaterial, of strategies and alliances, of institutional interventions, both secular and ecclesiastical. The names of the guilds who offered the *ceri* to the saint have remained and the festival of the *Gigli* each year attracts tens of thousands of tourists and visitors. It is a game of political position-taking, of negotiated forms of power relations which, naturally, in Leone's time, was played out in a different way.

---

35 The other saint much venerated in Nola is Felix. Leone tells the story (see bk. II, ch. 12) that manna came out of the tomb where he was buried and was gathered up. He lived about a century and a half before Paulinus.

36 Arjun Appadurai (1996) has described the past as a storehouse or archive from which we draw as we think best, according to our present needs.

5   In Conclusion: Extreme Recycling

Funeral lamentations are a classic example of those types of behaviour which, over the passage of time, have become associated with the populace, although for centuries all over Europe the elites also participated in this practice. For Alberto Cirese, whose remarks formed the starting point of this essay, and other commentators, the history of funeral laments is a clear illustration of the dynamism of cultural phenomena, their shifting attachments, never definitive and never permanent, to models of behaviour which are adopted by higher or lower social classes or even shared by all. As already mentioned, the Renaissance was one of the periods of European history, in Italy in particular, when circumstances, including the extraordinary geographical discoveries of the time, encouraged an increasing refinement of the intellectual tools whereby social and cultural differences could be analysed and expressed. Leone deals with a small corner of the world but he too exemplifies this process. Funeral rites in Nola also included ritual lamentation.

Burial took place in the church: wealthy families had private chapels built— or had purchased them—which contained a grave where corpses could be placed. The poor when they died were put in common graves which were also within the perimeter of consecrated sites. Those who died suddenly, without, that is, receiving the sacraments, were buried in a special area outside the bishop's residence. When a person died, the church bells announced the death, with a different chime according to whether the deceased, distinguished or not, was male or female or a child, in such a way that everyone in the city would know who had just passed away. The deceased would lie at home for two days. The funeral was carried out by lay sextons (*vespillones*) belonging to a confraternity (*fratraria*), which brought them together in a bond of mutual solidarity and which had a church of its own and the responsibility for burying the dead.

The body was placed on a hearse to be carried out of the house. The women of the household weep and the widow cuts off her hair and scatters it over the dead man, embracing him.

> The women, weeping, with their hair loose and dressed in dark clothes, follow the corpse until it is placed on the bier. If the wife of the dead man is among them she cuts her hair in the presence of all and covers the chest of the dead man with it, as a gift, so that he carries his wife's hair into the tomb with him as her dearest part. After she has given him a final kiss and uttered loud lamentations, she is pulled away from the corpse she holds in her arms. Those officiating at the funeral and her husband's friends

lead her off as she weeps. Then the daughters and other female relatives, they too weeping, kiss the dead man, after which the bier is hoisted on to the shoulders of four sextons, young and robust, and carried away.[37]

The women are determined and hold on to the hearse, trying to follow it, but they are led back into the house by friends who have also arrived. In church the exequies are performed with greater or lesser solemnity according to the status of the deceased. The ceremony can include the construction of a platform, the burial in a personal tomb or a carved stone plaque. After the burial, the family of the deceased return home, where they find a dining table which has been prepared by their neighbours and friends. They sit down and eat, in search of relief and consolation.[38] A period of mourning ensues, during which men for several months wear a tunic and cover their heads, refrain from shaving and cutting their hair, and maintain a serious and calm composure; for women, it is worse, especially if they are near relatives of the deceased, his wife or daughters, since they are obliged to wear black dresses and veils perhaps for the rest of their lives, unless they marry.

Let us take a further look at the burial places. Leone explains clearly that private chapels need to be endowed with the income from lands or property, income which goes to pay the priests who are delegated to say prayers of suffrage at fixed times of the day. The chapels can be sold on, in which case the new purchasers empty the grave of the bones of the previous owners. At this point Leone gives us a startling and fascinating piece of information: "These [bones] are carried out of the city on carts as though for a second funeral and are scattered along the muddy roads so that passers-by will trample them and the roads become more dry".[39] The bones are placed on hearses as though for a second funeral and they are scattered on the muddy roads so that they are ground to pieces under the feet of passers-by, thus drying the road surface.

---

37   Leone 1514, bk. III, ch. 2, f. xxxxiii *recto*: "Secuntur ipsum mulieres usque ad capulum passis capillis, lugubri veste plorantes plangentesque, inter quas siqua uxor fuerit defuncti, ipsa palam resectis forfice capillis omnibus pectus mariti contegit ac relinquit largiens ut illos in sepulturam rem suam uxoris carissimam conferat, deinde osculis postremis datis lamentationibusque magnis factis evellitur a cadavere, quod amplexa tenet; plorantem illam evellunt magistri pompae ac viri amici. Postea filiolae atque affines mulierem etiam plorantes defunctum osculantur, atque ultimis his officiis actis tollitur capulum humeris quattuor validorum iuvenum vespillonum atque effertur".

38   On this subject see a classic of anthropological studies: de Martino 1983.

39   Leone 1514, bk. III, ch. 2, f. xxxxiii *verso*: "plaustris extra urbem tamquam alteris exequiis efferuntur atque in viis caenosis sparguntur ut viatoribus calcata reddant itinera sicciora".

There have been various methods of recycling human bones but this must be one of the most useful.[40]

But Leone's account of this ceremony is not yet finished:

> In the same way, the rectors of the churches, when they clean out the common graves or dig new ones, throw all the dirt and bones of the deceased mixed up with the soil out of the city and spread it under the feet of passers by, so that, as the human condition dictates, there is no place of rest not even for the dead and so that men realise there is no difference between themselves and the earth we tread with our feet.[41]

Thus, in the same way, the rectors ("praefecti") of the churches clean out the communal graves and throw everything they contain—rubbish, soil, bones—about the streets, as refuse to repair the road-surface. Leone even finds a moral in the practice, commenting that this is the human condition: there is no peace for our bones and everyone who walks on them will realise that they themselves are no more than earth to be trodden under foot. On this occasion, the evocation of some edifying episode in the remote past will not help Leone or us. Furthermore, the bones of the rich as well as of the poor are scattered in this fashion: under the feet of passers-by social and cultural differences finally count for nothing.

---

40   The Nazis carried out a similar operation to recycle bodies in various extermination camps: bodily fat was used to make soap, hair to make felt, while the bones were used as fertilizer. Human remains have often been put to new and more or less dignified uses: sacred relics which are venerated in Christian churches or the display of entire bodies or body parts in historical and scientific museums. The still living organs of the just dead can be transplanted. In exotic cultures objects and tools are made out of human bones, hair, teeth (see Favole 2003). See also, in the vast field of studies on thanatology, the essays collected in Remotti (ed.) 2006.

41   Leone 1514, bk. III, ch. 2, f. xxxxiii *verso*: "Prefecti quoque basilicarum vel qum publicas foveas mundare purgareque volunt aut novas aliquas excavare purgamina et terrae immista ossa defunctorum eodem modo extra urbem eiciunt, viatoribusque sternunt, ne ita ferente conditione humana sit aliquis etiam mortuis quieti locus, utque homo animadvertat inter ipsum atque terram, quam pedibus calcat, nihil interesse".

CHAPTER 8

A Bibliographical Note on Ambrogio Leone's *De Nola* (1514)

Stephen Parkin

DE NOLA / Opusculum. / Distinctum / Plenum / Clarum / Doctum / Pulcrum [sic] / Verum / Graue / Varium & / Vtile.

Collation: 2°: a–i6 k⁴ A–B⁴ [$½ signed, signed in roman numerals] (66 leaves), [i] ii–x [xi] xii–lviii, [8] (xxvii misnumbered xxviii). 4 engraved plates inserted after a3, b4, d5, e3 (London, British Library copy, 178.g.14).

Fingerprint: e-i. len. aqVe nuab (3) 1514 (R)

Contents:
a1ʳ: titlepage.
a1ᵛ: blank.
a2ʳ–a3ʳ: AMBROSII.LEONIS.IN.LIBELLOS.DE.NOLA. / PATRIA.AD ENRI CVM.VR- / SINVM.PRINCIPEM.IV- / STISSIMVM.
a3ᵛ: blank.
a4ʳ–d3ᵛ (b5ʳ, d5ᵛ blank): LIBER PRIMVS.
d4ʳ–g4ᵛ (e3ᵛ blank): DE.NOLA. LIBER SECVNDVS.
g5ʳ–k4ᵛ: DE.NOLA.LIBER.TERTIVS.
A1ʳ–B2v: INDEX.OMNIVM.EORVM:QVAE.PASSIM.IN. / HOC.OPER.CON TINENTVR.
B3ʳ–B4ʳ: CORRIGENDA.IN.HOC.OPERE.SVNT.HAEC.
B4r: [Colophon]: ¶Incussum est hoc opus opera diligentiaq[u]e Probi uiri Ioannis Rubri / Vercellani. Venetiis Anno Salutis. M.D.XIIII. Septembris uero die IIII.sub Leonardo Lauredano Duce Sapientissimo. / [Register]: Registrum. / a. b. c. d. e. f. g. h. i. Omnes sunt ternio[n]es. k. A. B. uero duernio[n]es. / [Privilege]: ¶Vnusquisq[ue] monitus sit:hoc opus impressum esse cum priuilegio & /gratia per decennium concessa ab illustrissimo Dominatu Ve- / neto. Nemo audeat igitur aliquo pacto ipsum imprime / re: si mulctam cupiuerit euitare: quae con- / tumacibus illic est indicta.
B4ᵛ: blank.

© KONINKLIJKE BRILL NV, LEIDEN, 2018 | DOI:10.1163/9789004375789_010

Type: roman (a5^v = 41 lines ; 20 lines = 110 mm).

References: EDIT 16: CNCE 45515

Our knowledge of sixteenth-century printing and publishing has vastly increased in recent decades, perhaps especially so in Italy, one of the coun-
tries where the new technology of printing and the subsequent commercial
developments in publishing and bookselling were most significant, above
all in Venice, which was for much of the period the undisputed centre of the
European booktrade. Through advances in analytical bibliography we know
much more about how books were actually produced on the printing press
and the technical and editorial interventions which were made in the course
of production while archival and other historical research has expanded our
sense of authors' relations with printers and publishers and how their books
were commercially protected, were controlled and censored by State and
Church, were acquired, read and collected by different kinds of reading (or listening) publics. Yet what we know is still very much the tip of an iceberg; much
remains mysterious and many editions, even celebrated ones, are still deeply
puzzling for a variety of motives—what occasioned them, what their purpose
was, how they were received.[1]

Ambrogio Leone's *De Nola* is a good example. What was Leone's purpose
in writing it and when did he compose the text? Why did he choose Giovanni
Rosso (Johannes Rubeus, who, in his imprints, customarily added 'Vercellensis'
or 'da Vercelli' to his name, after his birthplace) as his printer and how closely
involved was he as the author with its production, its format, its visual design
and presentation? How well did the book sell and to what kinds of readers?
How many copies were printed and how many copies survive? Some of the essays in the present volume have provided convincing answers to some of these
questions: Fernando Loffredo has shown how the Venetian renown of Niccolò
Orsini, the penultimate Count of Nola, who fought on the Venetian side in the
War of the League of Cambrai, may well have created and triggered the commercial opportunity for the publication of a work on a subject, Leone's native
city in the far south of the peninsula, which would otherwise have been, in
Venetian eyes, obscure and marginal, while Fulvio Lenzo's analysis of the four
engravings which are part of the edition reveals, in terms of their content and
their production, how closely Leone must have been involved with the actual
printing of the volume.

---

1   I am grateful to Prof. Neil Harris of the University of Udine for bibliographical help in writing
     this piece.

The availability of bibliographical databases, with their increasing comprehensiveness, enables us to track more easily than ever before surviving copies of Leone's book and trace their current geographical distribution, at least in public or institutional ownership. As might be expected, the majority of copies—twenty-three—are held in Italian libraries, as the *Censimento nazionale delle edizioni italiane del XVI secolo* makes clear;[2] rather fewer copies have made their way outside Italy, at least into public collections: fifteen, of which five are held in North American libraries and ten in Europe, four of which in England. A more detailed and analytical census of the copies based on this data might serve to show significant bibliographical variants relating to the actual printing of the edition (one important difference which has already been noted, though not systematically mapped, is the variations in colouring in the engravings) as well as uncover details of past ownership and patterns of movement which would reveal something about the contemporary (and subsequent) acquisition and reputation of Leone's work.

The reasons for Leone's choice of Giovanni Rosso as his printer must, in the absence of archival evidence such as contracts and correspondence, remain even more speculative. It has been seen as curious that Leone did not publish his work with the leading scholarly printer-publisher of the period, Aldus Manutius, with whom he is known to have been on familiar terms.[3] There could be various explanations: Aldus's publishing programme was highly structured—this is one of the celebrated aspects of his intellectual (and commercial) achievement—and Leone's work would have represented a departure from this, not least, as the essays in this volume have shown, in terms of its originality of approach and experiments with genre. Nor was Aldus interested in illustrated books (the extraordinary achievement of the *Hypnerotomachia Poliphili* is the exception which proves the rule) and there is clear evidence from Leone's work that he saw the four images of Nola as central to his project. Moreover, although Aldine editions were often the fruit of scholarly collaboration between Aldus and the editors who worked with him, he—together with his marked emphasis on the uniform visual appearance and presentation of

---

2   At least two further known copies should be added to this total: a second copy held by the Biblioteca Nazionale Vittorio Emanuele III in Naples (as Andrea Ruggiero points out, Leone [ed. Ruggiero] 1997, 19–20) and a copy in Capua, Biblioteca Arcivescovile (listed by Pietro Manzi in Manzi 1974, 66). It is highly probable that further copies survive, both in Italy and elsewhere, which are unrecorded in current databases, to be found either in private collections or as yet uncatalogued (or at least uncommunicated to databases) in institutional libraries.

3   For example, by Andrea Ruggiero (1997, 18, n. 28).

his editions—may have been less welcoming to the kind of authorial control Leone appears to have exercised over this work in praise of his birthplace and the form in which it emerged from the printing press. Instead, Leone chose to work with Giovanni Rosso (Johannes Rubeus), at the time of the publication of the *De Nola* the longest established printer in Venice, who by the time of his last edition in 1519 had accumulated forty years of more or less continuous activity. Rosso had started printing in Treviso in the early 1480s but soon moved to the much larger centre of Venice; his first recorded Venetian production appears to be the edition of Josephus, *De antiquitate Judaica* printed in October 1486. He concentrated predominantly on Latin works, of classical and humanist texts (Suetonius, Livy, Pliny, Quintilian, Platina, Guarino of Verona, Filelfo, etc.), with rare excursions into the vernacular (although these included three important large-scale editions, perhaps significantly financed by others: in 1497, "ad instantia del nobile homo miser Lucantonio zonta [Giunta] fiorentino" (the founder of the famous dynasty of publishers, printers and booksellers) the Bonsignore translation of Ovid's *Metamorphoses*, illustrated by a celebrated series of large woodcuts, and two editions, in 1487 and 1494 respectively, of the Niccolò Malermi translation of the Bible (the later edition again for Lucantonio Giunta). With the new century Rosso's activity became more sporadic (24 recorded editions between 1501 and 1519, as opposed to *circa* 68 from 1479 to the end of the fifteenth century) but the three years from 1512 to 1514, the year of *De Nola*, saw a small flurry of production, including, unusually for Rosso, new works by contemporary authors: Jacopo Sannazaro (an edition of the *Arcadia*), the *De modo scribendi* by the Ferrarese mathematician Sigismondo Fanti (a treatise on calligraphy, in Italian, despite its title) and the *De Nola* by Leone himself. This cluster might suggest an opening up of commercial horizons, for whatever reason and however short-lived, for Rosso which could in itself have attracted Leone to the printer. Compared with the contemporary and highly fashionable Aldine editions (a direct contrast can be made between Rosso's rather cumbersome quarto *Arcadia* of 1512 and the Aldine octavo edition two years later, dedicated, as between peers, by Aldus to Sannazaro, with the proud but obliquely stated boast that the poet's learned and elegant work, now printed by Aldus, "facta est etiam mea"), the appearance of the printing, while decorous, is decidedly old-fashioned but working with Rosso may have allowed Leone more scope to supervise the project and produce the kind of book he envisaged. Leone's interventions in the edition are evident from the outset with the highly idiosyncratic title he gave the work, its somewhat quizzical terms printed as a vertically stacked list in the centre of an otherwise empty folio title-page, but they are perhaps most notable with the inclusion of the four engraved plates.

The presence of these plates has been the main reason for the book's bibliographical fame (and collectability), since they are one of the first uses of engravings as book illustrations in the history of printing and publishing (and the first such use in Venice). The field of book illustration had been dominated until then, and would be for several decades to come, by woodcuts, which had also attained, especially in Venice, an outstanding level of technical and artistic excellence (it is enough to think of the remarkable woodcuts, created in Venice, for the publication in 1543 in Basel of Vesalius's *De humani corporis fabrica*). There is nothing in the engravings to the *De Nola* which could not have been rendered as well if not better as woodcut images so it seems a plausible hypothesis that Leone chose to use engravings out of a wish to experiment with new techniques, as a pioneering innovation, the essentially trial status of which their indifferent artistic quality would seem to bear out,[4] but one which was in accordance with the quite conscious new-fangledness of the work as a whole, also in terms of its content.[5] The already mentioned intermittent use of different colouring in the four engravings recorded in many of the surviving copies would also seem to reflect this experimentalism, while the sequence of pages left blank in the imposition of the sheets (a3^v, b5^r, d5^v, e3^v) would appear to be evidence of some technical uncertainty on the part of Rosso and his printers over how to combine intaglio printing with letterpress. Was the initial—farfetched—intention to try to print the engraved images onto these pages? The difficulties which arose meant that the expedient, which of course became the standard technique with book engravings, of

---

4   See, for example, the great collector Philip Hofer's remarks in his essay on the early use of engravings as book illustrations (Hofer 1934, 301): "[The *De Nola*] is generally rather a shabby book, with four indifferently printed illustrations which are hardly more than rough maps". The detailed analysis by Lenzo in the present volume, 59–80, of the images and their complex relation to Leone's text shows how hasty Hofer was in judging these illustrations to be little more than "rough maps". Hofer's copy of the *De Nola* was given with the rest of his extraordinary collection of illustrated books to the Houghton Library of Harvard University (it is the copy described in Mortimer 1974).

5   See the Introduction to the present volume, 1–10, on the "profound originality" of Leone's text. Lorenzo Miletti in his essay on the *De Nola* as a Renaissance work, 11–40, notes Leone's unusual and, given the book's length and substance, seemingly inappropriate use of the term 'opusculum' in the title: perhaps the author's choice of this curious diminutive derives in part from his sense of the work as a kind of "essay" (in the—later—Montaignian sense of 'trial' or 'experiment').

the binder inserting the separate leaves of plates into the text gatherings had to be adopted.[6]

An extended bibliographical analysis of the first edition of the *De Nola*, towards which the present note serves merely as a preliminary paragraph, would thus appear to confirm and reinforce our sense, found throughout the present collection of essays, of Leone's work as a compellingly idiosyncratic achievement both intellectually, in terms of its content, and materially, as a printed book.

---

6   On each of the blank pages in the British Library there is a note, in a sixteenth-century hand: "h[a]ec cu[m] sequente e[st] incollanda", which may suggest another and more plausible hypothesis that the engravings were intended to be stuck down on the blank pages (rather than bound in alongside, as they have been in the actual copy). From the available catalogue descriptions of copies held elsewhere, there is no indication that this solution has ever been adopted. For a discussion of the difficulties of combining the processes of intaglio and letterpress printing, see Gaskell 2004.

Appendix of Texts

ed. by Lorenzo Miletti

1 *De Nola*'s Table of Contents

LIBER PRIMVS

LIBER SECVNDVS

© KONINKLIJKE BRILL NV, LEIDEN, 2018 | DOI:10.1163/9789004375789_011

10. De porticu Nolana atque viis urbis (f. xxx *recto*)
11. De basilica urbis principe et episcopio (f. xxxi *recto*)
12. De basilica et episcopio antiquo deque reliquis [sic] divorum quae inibi erant (f. xxxiii *verso*)
13. De reliquis urbis basilicis (f. xxxvi *recto*)
14. Qua ratione in urbe praesente sint subterranea cavave sacella (f. xxxviii *recto*)
15. Quae sit figura aedium praesentis urbis et qualiter earum partes se habeant (f. xxxviii *recto*)
16. Qui fuerint qui aedificando benemeriti sunt de urbe Nola (f. xxxix *verso*)
17. Quae fora et quibus in locis pateant atque quae res in eis tractent (f. xxxx *verso*)

LIBER TERTIVS

1. De sacris initiatis et dignitatibus ac proventu eorum item de beginis et vespillonibus Nolanis (f. xxxxi *recto*)
2. De sacellis sepulcrisque privis [sic] deque funeribus atque eis quae circa funera tractantur a Nolanis (f. xxxxii *recto*)
3. De familiis praesentis urbis egregiis (xxxxiiii *recto*)
4. De Iudaeis qui Nolae saeculo nostro habitavere ac eorum eiectione (f. xxxxix *recto*)
5. De studiis Nolanorum praesentium (f. xxxxix *verso*)
6. De elegantia et gratis moribus Nolanorum (f. l *verso*)
7. De magistratibus Nolanis (f. li *verso*)
8. De tributis et vectigalibus quae in urbe populoque Nolano fiunt (f. liii *recto*)
9. De matrimoniis et dotibus (f. liii *verso*)
10. De spectaculis histrionum atque ludo equestri (f. liiii *recto*)
11. De consuetudine quadam sacerdotum praesulem venerantium (f. lv *verso*)
12. De altera sacerdotum consuetudine (f. lvi *recto*)
13. De more quodam adolescentium Nolanorum (f. lvii *recto*)
14. De vindemia et more quodam paganorum qum vindemiam faciunt (f. lviii *recto*)

2   *De Nola, Praefatio* (f. ii *recto*–iii *recto*)

AMBROSII LEONIS IN LIBELLOS DE NOLA | PATRIA AD ENRICUM UR|SINUM PRINCIPEM IU|STISSIMUM

1. Omnes homines, Enrice Princeps, vehementer optant mirificeque nituntur ut patriam tam ope summa iuvent quam omni officio, veneratione ac pietate prosequantur. Cari namque sunt filii, cari parentes, cari quinetiam atque nihilominus amici, omnium

tamen patria una longe carissima est. Non solum enim patria parentum esse parens, atque munera quae a parentibus proficiscuntur omnia contulisse videtur, verum etiam complura nobis alia comperitur praestitisse veluti solem, aethera, humum, orationem, documenta, mores, nomen, societates, domos atque alimenta. 2. Quibus muneribus simul se benignam atque suavem ostendit, simul etiam nos usque adeo secum coniunxit, ut siquid facto dictove sive alii sive nos fuerimus de ea benemerentes id aeque ac nobis ipsis accesserit existimemus atque summopere gaudeamus. 3. Eamobrem combustam dirutamque a Gallis Romam Furius Camillus ardenti animo, cura, sollicitudine ab interitu vendicavit, instauravit, adauxit. Alfonsus item rex Aragoneus secundus, Enrici avi tui frater, Parthenopen patriam, quae iampridem aperta prope atque angusta fuerat, tum plurimum amplificavit tum moenibus quam praestantissimis circundedit. 4. Athenas quoque alioquin inter urbes illustris facile clarissimas Isocratem ac Aristidem inter primos laudibus magnis iuvit extollere, idque dicatis Panathenaicis orationibus luculentissimis. Innumeri quinetiam comperti sunt viri qui se corporaque pro vallo ausi sunt hostibus obiicere eligentes mortem cum vita commutare potius quam sinere patriam vinci aut vastari aut laedi. Alii insuper aliter operam egregiam navavere pro patria quos recense- | (ii^v) -re opus foret immensum.

5. Ipse vero patriam quoque studio prosecuturus, ut horum omnium bene faciendi cupiditatem videor imitari, ita modo quodam sum dispar munere ipso reddendo. Illi enim patriam opere magno redintegravere vel effecere maiorem, ipse imagine atque aspectus proprii similitudine tanta produxi in medium, adiutus opera Hieronymi Moceti pictoris, ut oculis omnium atque ubique terrarum perquam facile possit esse conspicua. Quinetiam illi patriam concinne affatimque laudaverunt ac evexerunt ad astra, ipse contra copia dicendi seposita relictisque floribus et orationis lenociniis sola sum historiae simplicitate complexus. 6. Imagine namque pura atque historia defaecata in hoc opusculo decrevi patriam venerari. Imaginis enim vis atque actio divina nimirum est, quippe quae veram suamque rei cuiusque faciem agit et offert atque absentem tam exacte praesentem ostendit, ac si foret in conspectu. In historia vero non ipsa res solum universa et vires ipsius et mores et honor et dignitas et gesta lucent, verum etiam eius inesse comperitur promissio magna quaedam aeternitatis. 7. Videor quinimmo interim operae pretium facturus tam civibus Nolanis quam hominibus exteris: hi enim nomen Nola audientes in quadam inani fictaque cogitatione tamquam insomnio vagantur, spectantes vero haec nostra, urbem situmque eius atque plurima alia quae sunt cognitione digna, aeque ac si praesentia forent infallaci certaque deprehensione novisse valebunt. 8. Cives vero, qui hac tempestate sunt, praesertim qui non multa callent ut hodiernam tuentur, ita veterem Nolam persentiscere nequeunt, etiam si nomen et gesta multa urbis audiant, simulatque haec ipsa, quae nunc edidimus, didicerint, antiqua cum praesentibus coniungentes, rem integram exquisite norint. 9. Posteris vero civibus non modo utilitas memorata provenerit, sed etiam praesentem urbem luculenter descriptam capient a nobis, quam nos ne tenebricoso quidem verbo significatam ab aliquo eorum qui ante nos fuere accaepimus.

Illo quoque posteri Nolani iuvabuntur, siquidem cernere poterunt quo pacto cum iis qui nunc sunt in urbe se habeant: ut novi aut recepticiine et ex collecta illuvie cives an indigenae sint, ac ex obscurisne ad claras familias surrexerint an ab optimatum genere ad ignobiles abiectosque degenerarint. 10. Et praesentis urbis tam figuram quam vires ac morum ordines, item tam urbanica quam agrestia conferre quam facillime poterunt cum eis quae suae cuiusque tempestatis fuerint num eadem ipsa manserint an deleta mutatave extiterint. Non enim esse solet idem rerum vultus hoc atque altero curriculo temporis. 11. Utrisque praeterea plurimum conducet haec tractatio. Aestimantes enim sese atque patriae dignitatem liberiore sane animo consuetudinem iniverint cum externis, conscientia enim obscurae patriae magno impedimento loquendi libertati esse videtur. 12. Contra vero externi homines, qum Nolam non ignorarint magnam dignamque ac illius antiquae avitaeque virtutis prolem ac | (iii^r) haeredem extare, ut ratione locoque maiore excipiant ac habeant Nolanos fas est animum inducant. In urbe enim digna aeque cives egregie libereque educari ac emendatis moribus ac praeceptis institui reputantur. 13. Indubitatum etiam est nobilitatem patriae quam maximam fore civibus dehortationem atque cohortationem. Qui enim antiquam patriae dignitatem honoremque agendis in rebus animo servant, ii tam turpia et patriae indecora patrare perhorrescunt quam ad honesta gerenda ac suis digna municipibus accenduntur. 14. Verumenimvero, magnanime Princeps, quod ab avis atavisque tantae urbis imperium suscaepisti quodque urbs civitasque ipsa ob singulares ac regias virtutes tuas intenta erga te est teque amat atque colit, nefas est ut urbis imago haec, quam modo descripsimus vel aliorum sit, vel aliter quam urbs ipsa se gerat. Proinde eam cape vel ut tuam, vel ut tibi nuncupatam a nobis proque Nola una duas habe, alteram quam regas, alteram quam legas. 15. Divisum autem opusculum est in libellos tris, in quorum postremo praesentis urbis ac nostri temporis cives eorumque mores et consuetudines explicantur. In secundo vero ager atque ea quae ad structuram praesentis urbis spectant continentur. Atque in primo de eis quae ad urbem antiquam referuntur disputatur, qui ab agro hoc pacto auspicium sumpsit.

3   *De Nola*, bk. 11, ch. 15 (**xxxviii** *recto*–**xxxix** *verso*)

1. Qua figura quove modo antiquorum aedes habuerint Nolanorum, nulla earum inpraesentiarum extante, ignoramus. Ne igitur posteritas aliquando idem desideret cognoscereque nequeat, placet hoc in loco illas illarumque partes scribendo repraesentare. 2. Aedium itaque species aut quatriquetra conspicatur aut, quod frequen- (xxxviii^v) -tissimum est, oblongior quam lata patet. Aedium quinetiam ita habentium quaedam ac perquam paucae inveniuntur quae carent cavaedio, quaedam vero atque plurimae reperiuntur cavaedium continentes. Quae vero cavaedium amplectuntur, illae frequentius unum latus eius attingunt; multae tamen sunt quae binis illud lateribus

ambiunt, quaedam quae ternis, rarissimae omnium offenduntur quae undequaque cavaedium cingant. 3. Domus autem partes aediumve quae magno interstitio separantur,
sunt hae: atrium, cella vinaria, equile, conclave, cubiculum, porticus superior, bibliotheca, penus, navis, turris, cavaedium atque horti. 4. Atrium locus est in solo ipso
sedibusque domus, in quem primum per ianuam de via excipiuntur intrantes. Hoc sa-
epius protenditur adusque cavaedium, ut lucidius sit atque ad introducenda utensilia
expeditius. Idem latitudinis minimae est decem pedum, maximae vero xxv, altitudinis
vero xvi aut plurium ita ut usque ad xxxii pedes interdum tollatur quousque contigna-
tionem primam subierit. Est autem latius quam ianua, haec enim ampla est pedibus
quinque aut pluribus; nam usque ad xii pedes dilatatur, alta vero tantundem cum se-
misse aut triente aut quadrante. 5. Contignatio vero atrii frequentissime ex tignis constat, raro testudine tegitur; solum vero pavimento stratum est, quod ex calce et parvis
pumicibus, quos lapillos vocant, ac pauca arena conficitur, quod lithostratum seu pu-
micistratum volentes dicere Nolani astracum dicunt, verbum Graecum sequentes. 6.
Pumiculi vero hi tum in agro Nolano fodiuntur, qum altius fuerit excavatum, tum in
radicibus Gecalae collis occiduis, ubi quoque arena praestantissima abunde comperitur. Quae argumento sunt eorum collium imas sedes magna ex parte arenosas pumicu-
losasque esse, non argilleas atque aquarum sustentatrices. Atrium demum hoc
porticum vulgo appellant. 7. In eodem etiam solo equile, quam stallam vocant, aedifi-
catur; locus oblongius porrectus est, in quo equi mulique ad equitandum vehendum-
que curantur. 8. Est cella vinaria oblongior quoque, quod etiam cellarium, verbis illis
corruptis in unum hoc ipsum, vocant. In ea servantur vina doliis ligneis recondita; sunt
autem eiusmodi dolia vasa rotunda duabus partibus, inter quas porriguntur, compres-
sa duploque longiora quam lata; longa vero sunt quantum homo tendere potest bra-
chia, media in parte parum extumida, structura ex tabellis castaninis, interdum
quernis, tantundem longis, palmo latis, digito altis ad unguem commissis atque utrin-
que a plurimis circulis ligneis arctantibus perficitur. 9. Quin utraque frons aut cornus
dicas, orbe ex compactis asseribus tabulato clusa est, quem orbem tympanum appellant, quod speciem tympani praesefert; in media vero dolii longitudine foramen ad
amplitudinem mali terebratur per quod impletur vas; illudque cavonem nuncupant,
quod dolio fabricato cavatur. 10. Quinetiam in altero cornu tympanove tria duove fora-
mina parva faciunt, ut in unum eorum insito canaliculo vinum (ut lubet) hauriatur.
Hoc ipsum vas bottem vocant sectantes vocem Graecam, hoc est magnum concaepta-
culum, βου namque magnitudinem vult significare, θήκη vero id quod rem impositam
ambiendo ca- | (xxxix^r) -pit, ex quibus compositum est βουθήκη, butheca, a quo verbo
abscissa ultima remansit buthe et bothe et botte mutatione literarum atque pronun-
ciantis asperitate. 11. Atrio autem frequentius equile haeret quam cella vinaria: haec
enim strepitus odit, qui in atrio et in via fiunt saepissime. Super his autem aulis inferio-
ribus prima sternitur contignatio ex trabibus castaninis constructa; raro enim infernae
domus partes testudine operiuntur. 12. Equile quinimmo caret pavimento; cella vero

creberrimum pavimentatur atque in quibusdam cellis etiam fovea ad magnitudinem dolii fit, quae pavimento incrustatur. Hisce vero partibus cavaedium adiacet in quo plurimum puteus coronatur; is pluviarum aquarum nihil suscipit; nam ad eam quae suapte natura ac subterranea scatet, descendit. 13. Post cavaedium patent horti parvo dumtaxat pariete exclusi a cavaedio, interdum pars domus inter cavaedium hortosque interiacet. In eodem quinetiam solo frequentius et furnus et lavacrum, quod cantarum pannorum vocant, aedificatur. 14. Ab atrio praeterea cavaediove scalarum gradibus ascenditur supra primam contignationem, quod solarium primum nuncupant. In hac frequenter conclave et cubicula sociata struuntur. 15. Conclave autem vel Graece dicas triclinium, salam Nolani vocant a saltando. Nam in ea saltare solent vel qum nuptiae fiunt vel qum convivia parantur. Locus est longior duplo quam latus; in altum vero tollitur quantum est latus; solet imminere viae atque ab eo latere fenestris duabus magnis aperiri, unde quae fiant in via (ut lubet) spectarent. 16. Haerent conclavi cubicola, quas cameras appellant; haec laudatissima sunt quae exacte quadrata figura cernuntur, latera enim singula ad xxxii pedes extendi solent, singula autem cubicula singulis alterisve fenestris illustrantur; fenestrarum altitudo est sex pedum, latitudo trium. Quarum podium non altius tollitur quam ut ad cingulum nostrum pertingat. 17. Conclavis vero fenestrae multo sunt maiores quam cubiculorum. Interdum cubicula non omnia conclavi haerent, sed eorum unum solummodo, cui caetera deinceps sunt adiecta ab eoque ad reliqua transitur, quae aut serie diriguntur aut primo cuncta haerent cubiculo. Habet conclave caminum magnum cum foco; cubiculum vero haec eadem sed dimidio minora. 18. Haeret quoque conclavi bibliotheca. Locus est cubiculo minor, interdum aequalis; in eo libri in pluteis pendentibus parietive affixis panduntur; hic contemplatur literariisque rebus incumbitur tractaturque, quod studium vocant. 19. Interdum etiam haeret conclavi porticus superior; huius aut duo latera arcus sunt dumtaxat insidentes super pilis columellisve, aut unum latus adapertum est uno duobusve arcubus, quam logiettam, hoc est scaenulam, vocant. Scaenam enim logia<m> vocant, quod in ea sedentes sub umbra λόγον id est sermonem serere solent et colloqui. Tegi consuevit ea porticus contignatione laquearibus pulc<h>ris hornata, raro vero testudine picta; usus huius etiam est ad captandum lucidiorem amplioremque aeris prospectum atque auras excipiendas. 20. Aedificatur quoque in hac prima contignatione coquina. Illud tamen referre oportet: nam in quibusdam aedibus sunt coenationes solummodo, in plurimis vero coenacula et coenationes, ubi et amplitudo soli et animus magnus atque diviciae con- | (xxxix^v) -senserint quemadmodum ab horum contrariis contingit ut in aedibus plaeraque ex hisce desint. 21. Vicina quinetiam conclavi manet penuaria cella, quae saepius in coenatione statuitur, quam spensam vocant. Prima autem contignatio orditur ex trabibus castaninis passim super duobus parietibus sitis atque ex transverso asseribus coassatis confixisque atque his iniecto pavimento pumici strato. 22. Teguntur conclave, cubicula, caeteraque quae super contignatione erecta sunt, ab altera contignatione quae inferiori similis edi solet, interdum vero ex trabibus ac asseribus absque pavimentatione construitur, super qua plurimum unus atque indistinctus locus patet, qui super coenaculis exurgentes etiam habet parietes cum crebris ac parvis fenestris; cooperitur vero hoc supremum spatium a pectinato tecto, interdum a semifastigiato ac imbricibus iniurias coeli defendentibus. 23. Usui autem hic locus Nolanis est ad stipandas servandasque maxime avellanas, quarum ingens nunc hisce copia nascitur. Appellant hunc locum navem domus a similitudine longae navis, quae pectinato tegmento solet operiri. A Nolanis vero nuncupatur verbo Graeco magacenum avellanarum, μεγάκενον enim Latini interpraetantur magnum vacuum; est enim hic locus magnus vacuusque, ubi merces multae stipari atque servari possunt. 24. Turris autem rara in aedibus Nolanis extat, sicuti rara pavimenta subdialia; his enim aliqui cives loco tecti utuntur. At vero tertia contignatio omnium rarissima in aedibus Nolanis reperitur. 25. Fabricae autem Nolanae non latericiae, tophaceae sunt omnes: ex cementis enim topheis, arena, calceque corrupta conficiuntur. Numerus aedium Nolanarum omnium, quaeque in urbe nunc sunt, ad septies centum censetur. 26. Sunt insuper aedes eiusmodi semotae a tabernarum locis, quod factum est honestioris vitae causa. Non enim decens esse ducunt ut in eo loco matronae et puellae inhabitent in quo omnis est virorum consuetudo. Namque tabernae ac meritoriae domus sitae sunt in media urbe circa porticum, circa emporium, circa forum frumentarium, circa insulam basilicae; caeterum vero urbis solum aedes occupavere. 27. Quamobrem virorum omnis congressus et exercitium interdiu agitur in tabernis et in earum viis, noctu vero quisque ubi servavit tabernam suas ad aedes se confert. Ea vero loca quae supra tabernas manent aut megacena sunt non hominum habitationes, aut diversoriarum domorum triclinia cubiculaque, atque praetoris est noctu custodire tabernarum solitudinem circa eas assiduas excubias facientis.

4   *De Nola,* bk. III, ch. 3 (f. xxxxix *recto*)

1. In hac quoque familia Beatrix puella venustissima illuxit, cuius ora ne an ingenia potius divina forent non erat hominis etiam Paridis opus diiudicare: utraque enim visa sunt summa esse atque supra quam ubi vis humana constituta est evecta. 2. Hanc Thomas Malvicus, sculptor praestantissimus, e marmore finxit, sed inter coelandum creduntur Gratiae atque Horae affuisse, opificisque oculos illuminasse, manusque eruditissimas reddidisse; siquidem marmore Pario miranda proprietate edita est, ubi enim marmor vivaque coram praestoque fuissent, Beatrices geminae cernebantur: altera audiens, loquens altera; ridens altera, altera laeta. 3. Neque Malvicum solummodo magnus ille puellae decor atque splendor ad effingendum commovit; verum etiam omnes Italos poetas ad canendum excivit, atque inter primos Antonium Thebaldeum Ferrariensem, virum ut divino ingenio ac eloquio fulgentissimum ita viris doctissimis

assidue atque ubique terrarum laudatissimum. Quam ob rem poematis innumeris utraque illa divinitatis imago celebrata est, praenomenque accepit a Musis: namque diva Beatrix ad omnibus est cantata. 4. Eadem duodevicesimum iam annum agens atque ad Gecalae pulcherrima iuga profecta qum inter oleas quercusque ascenderet e medio, rem miram, statim tempore sublata est, nec usquam a quoquam deinde visa fuit. Creditur tamen ob eius optimos divinosque mores viva volare cum superis atque alto successisse coelo.

5   *De nobilitate rerum*, ch. 41, (f. g viii *verso*–h i *recto*)

1. Quin etiam nostra tempestate duorum Mediolanensium genere, sculptura clarissimorum, opera subtilissima et ingeniosissima florescunt atque praedicantur. Alteri Caradossus, alteri Daniel[1] ex familia Arcionum nomen est. 2. Verum hic in eo genere praesentim pollet quod niellum novato verbo appellant, neque aliter in eo splendet quod vitreum est, ipsi vero fusores smaltum vocant. Nam in hisce tanta subtilitate et ratione hic usus esse probatur, ut ab omnibus in ea arte claris summo honore habitus sit. Quid quod opera eius tam mira digestione iuncturaque et gratia gaudent ut antiquorum signa praeclarissima ad certamina convocare valeant? Inter quae[2] salinum confecisse constat ex xiiii unciis argenti (est autem uncia[3] pondus septenorum aureorum cum dimidio aurei) quod Romae invenit qui septicentis aureis emit. Nam aureus est nummus ex auro impressus, octoginta granorum tritici pondere. 3. Caradossus vero eductis imaginibus eminentibusque et sculptis nitidior, is quidem tum externorum, tum Italorum palmam tulit. Nam excellenti acumine vel artis, vel ingenii sua prolata sunt, quae a Praxitele Lysippove[4] an a recentiori efformata sint, vel periti vix profecto iudicabunt. 4. Exstat enim eius calamarium, ut cetera praeclara et nobilia omittamus, ubi omnes artis facultates censentur adesse, quodque ulterius progrediendi ingenio locum non reliquisse visum est. Nam in uno illius latere nudi equis insidentes spectantur, qui auxilio cuidam puero venerant, quem aquila eripuerat in coelum; illi vero suspicientes alitem puerum deportantem, eum animum propositumque ostendunt ut evolare cum equis quoque velle videantur, ubi eis figuris raptum Ganimedis ostendit. 5. In altero pugna Centaurorum cum Lapithis[5] est. Tertio latere Hercules est qui Cacum suppositum, laevaque gutture illius presso, et genu altero cum violentia quadam stomachum calcante, alteraque os stringente compulsat. Quarto vero Hercules leonem exossans spectatur, adeo pulch<r>e exculptus, ut hominem ira percitum, leonem dolore gementem prope sentiretis. Ex quibus lateribus, ob subtilitatem et operis excellentiam plurimae sulphureae tabellae fusae sunt, quibus per totam Italiam opus summa cum admiratione spectatum est. 6. Pro quo Ioannes Aragoneus, Ferdinandi filius, mille et quincentos aureos spopondisse fertur.

---

1 *Danielis* Leone; 2 *interque* Leone; 3 *unicia* Leone; 4 *Lisippove* Leone; 5 *Laphitis* Leone.

Illustration Section

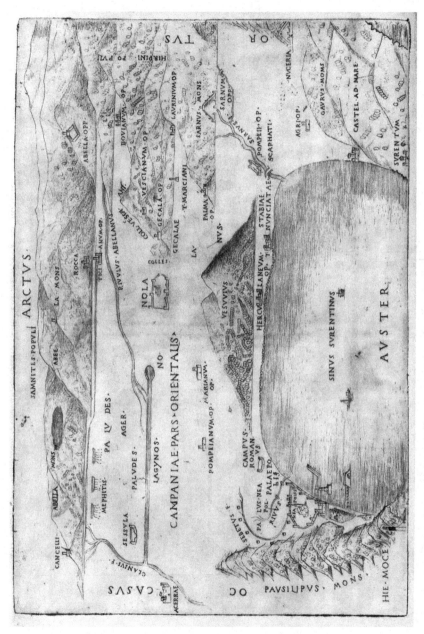

FIGURE 1   *Ambrogio Leone, Girolamo Mocetto, Ager Nolanus, engraving, first state.*

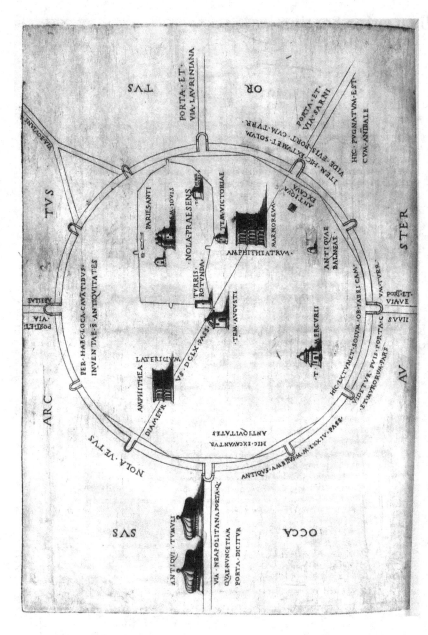

FIGURE 2   *Ambrogio Leone, Girolamo Mocetto, Nola Vetus, engraving, second state.*

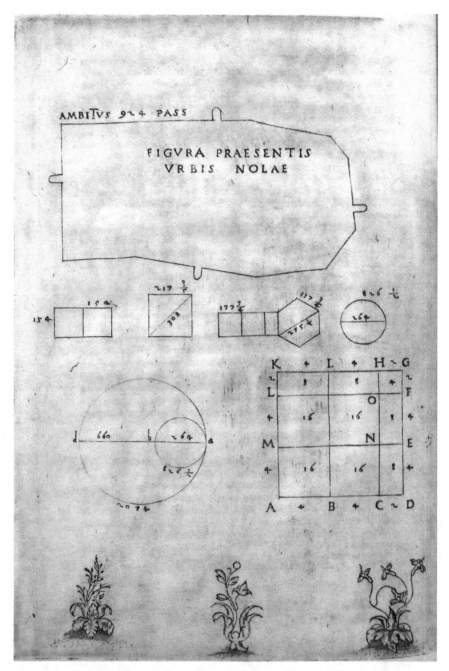

FIGURE 3   *Ambrogio Leone, Girolamo Mocetto,* Figura praesentis urbis Nolae, *engraving, second state.*

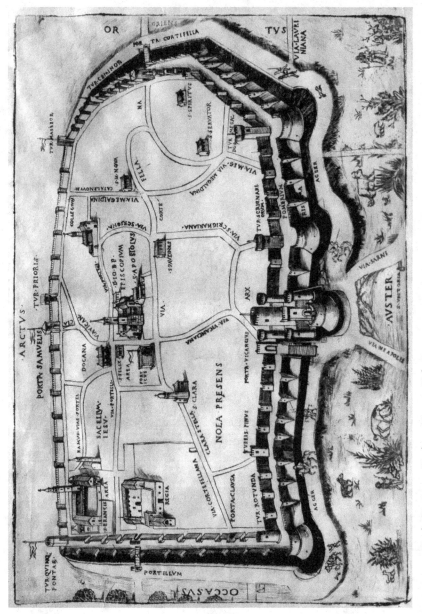

FIGURE 4   *Ambrogio Leone, Girolamo Mocetto, Nola Praesens, engraving.*

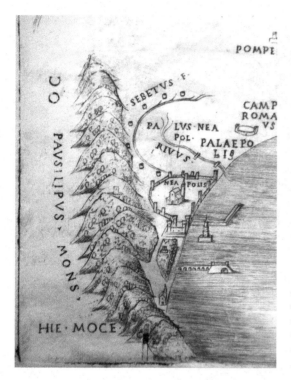

FIGURE 5
Ambrogio Leone, Girolamo
Mocetto, Ager Nolanus, *first*
state. Detail including the view
of Naples and, on the left, the
signature of HIE[ronimus]
MOCE[tus].

FIGURE 6
Ambrogio Leone, Girolamo
Mocetto, Ager Nolanus, *second*
state. Detail including the view
of Naples, the signature of
HIE[ronimus] MOCE[tus] and a
little plant.

FIGURE 7   *Naples, Castel Nuovo. Vault of the Gran sala, 1452–1458.*
              PHOTO FULVIO LENZO.

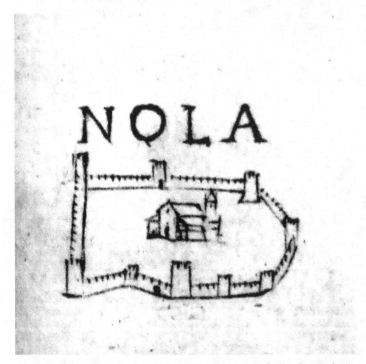

FIGURE 8   *Ambrogio Leone, Girolamo Mocetto,* Ager Nolanus. *Detail of Nola.*

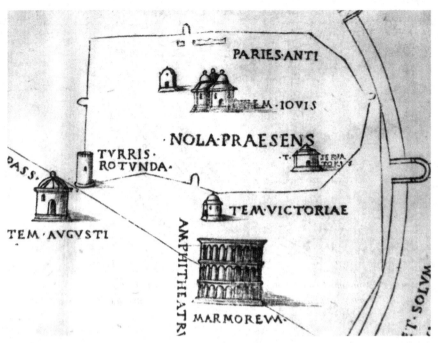

FIGURE 9   *Ambrogio Leone, Girolamo Mocetto, Nola Vetus. Detail representing the area of*
modern Nola within the larger perimeter of the ancient city.

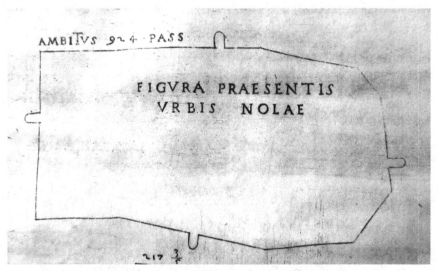

FIGURE 10   *Ambrogio Leone, Girolamo Mocetto,* Figura praesentis urbis Nolae. *Detail represen-*
ting the modern city.

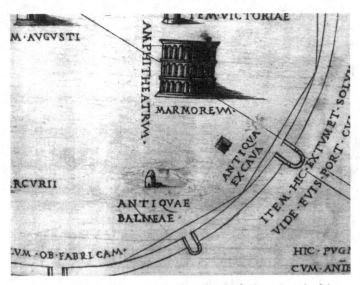

FIGURE 11   *Ambrogio Leone, Girolamo Mocetto,* Nola Vetus. *Details of the "marble amphitheatre", thermal building and of the "antiqua excava".*

FIGURE 12
Ambrogio Leone, Girolamo Mocetto, Nola Vetus. *Detail of the thermal building in another copy of the De Nola.*

FIGURE 13   *Avella, Pedestal bearing the inscription CIL, X, 1211. Limestone. View of the right flank showing a relief of an amphitheatre.*
PHOTO HISTANTARTSI ARCHIVE.

FIGURE 14   *Ambrogio Leone, Girolamo Mocetto,* Nola Vetus. *Detail of the temple of Mercury.*

FIGURE 15   *Matteo de' Pasti, Foundation medal of the church of San Francesco in Rimini, 1447–50. Bronze, diameter 4 cm.*

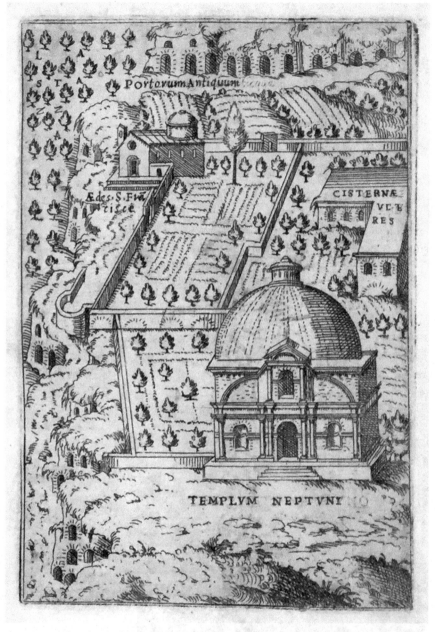

FIGURE 16 *Anonimous, Reconstruction of the Temple of Neptun in Pozzuoli. Engraving, after Francisco Villamena*, Ager Puteolanus sive Prospectus eiusdem Insigniores, Rome *1620, plate 10.*

FIGURE 17   *Anonimous draftman, Map of the "Selva di Maddaloni".*
Detail of river Clanio and and Virgil's verse "Clanius
non aequus Acerris". Naples, Archivio di Stato, Ufficio
Iconografico, *65. Ink on parchment.*

FIGURE 18   *Anonimous draftman, Map of the Selva di Maddaloni.*
Detail of Nola and the Roman amphitheatre identified
as "antic. anphit.". Naples, Archivio di Stato, Ufficio
Iconografico, *65. Ink on parchment.*

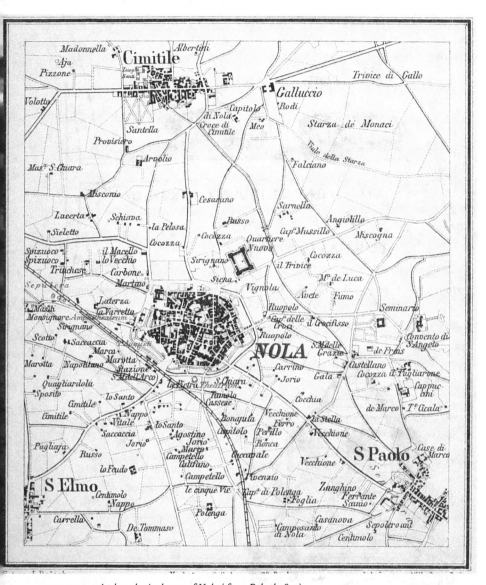

FIGURE 19   *Archaeological map of Nola (from Beloch 1890).*

FIGURE 20   *Ambrogio Leone, Girolamo Mocetto*, Nola Vetus. *Detail of the two tombs* (antiqui tumuli).

FIGURE 21
Anonimous draftman (Destailleur B),
View of an ancient tomb from Nola.
Saint Petersbourg, Ermitage, codex
Destailleur B, f. 110(107) verso, detail.
Ink on paper (cm 15,3 × 20,6).

FIGURE 22   *Antoine Morillon, Ancient funerary monument in Nola. Eton, Eton College, ms. Bo.17.4, n.4), f. 12 recto: "Nol[a]e ex Albano lapide". Ink, wash on paper.*

FIGURE 23   *Antoine Morillon, Ancient funerary monument monumento in "Tegiani". Eton, Eton*
            College, ms. Bo.17.4, n.4. Ink and wash on paper.

FIGURE 24 *Avella, Roman mausoleums.*
PHOTO HISTANTARTSI ARCHIVE.

FIGURE 25 *Francesco di Giorgio, sketch of three ancient monuments "in champagna". Florence, Uffizi, Gabinetto Disegni e Stampe, 337 A recto, detail.*

FIGURE 26   *Nola, Antiquarium of the Archdiocese. Inscription CIL, X, 1272. Limestone.*
Height: cm 72,6. Width: cm 123.
PHOTO HISTANTARTSI ARCHIVE.

FIGURE 27   *Nola, Antiquarium of the Archdiocese. Inscription CIL, X, 1285. Limestone.*
Height: cm 19. Width: cm 85,5.
PHOTO HISTANTARTSI ARCHIVE.

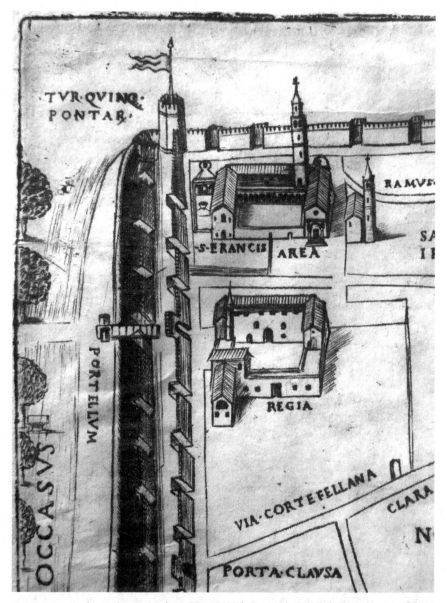

FIGURE 28   *Ambrogio Leone, Girolamo Mocetto,* Nola Praesens. *Detail showing the area of the Orsini palace* (regia) *and the church of San Francesco.*

FIGURE 29   *Nola, Orsini palace. View of the facade. 1470.*
                PHOTO HISTANTARTSI ARCHIVE.

FIGURE 30   *Nola, Albertini palace, Detail of the base with spolia, c. 1470.*
PHOTO HISTANTARTSI ARCHIVE.

FIGURE 31   *Naples, Collegium of the Gesù Vecchio (once the palace of Carlo Carafa).*
Detail of the base executed with ancient white limestone blocks from the
remains of the "Temple of Mercury" in Nola.
PHOTO FULVIO LENZO.

FIGURE 32
Nola, Albertini palace.
Courtyard. c. 1470–1530.
PHOTO HISTANTARTSI
ARCHIVE.

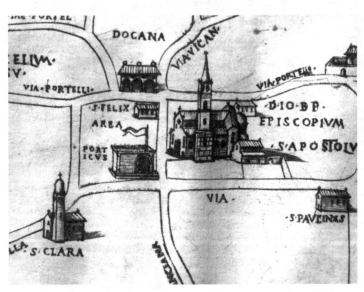

FIGURE 33   *Ambrogio Leone, Girolamo Mocetto, Nola Praesens. Detail showing*
the cathedral, the seggio, the Dogana and the churches of Santa
Chiara, San Felice, Santi Apostoli and San Giovanni Battista.

FIGURE 34   *Master of the Bardi Chapel, Bardi Altar, Detail showing the Choice of the habit in the Bardi dossal, Church of Santa Croce, Florence, 1263–1266.*
© SCALA ARCHIVE.

FIGURE 35   *Cimitile, church of San Felice.*
PHOTO HISTANTARTSI ARCHIVE.

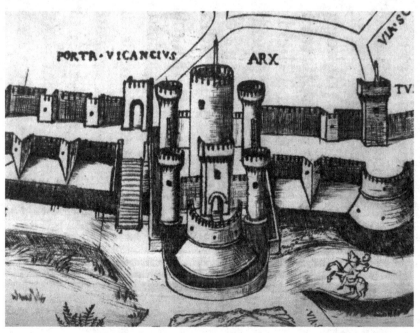

FIGURE 36   *Ambrogio Leone, Girolamo Mocetto,* Nola Praesens. *Detail showing the* Arx
(the castle).

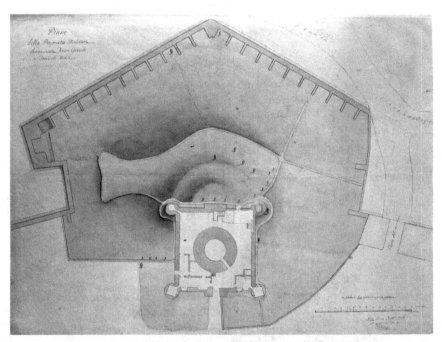

FIGURE 37   *Anonimous draftman*, Pianta del Piano di Proprietà militare denominato Fosso
Grande o Torre di Nola, *1816. Naples, Biblioteca Nazionale, Manoscritti e Rari,
b.5.C.95. Pencil, ink, wash on paper.*

FIGURE 38   *Vincenzo Franceschini, View of the castle of Nola. Oil on canvas (cm 25 × 39). Naples
Museo Nazionale di San Martino.*

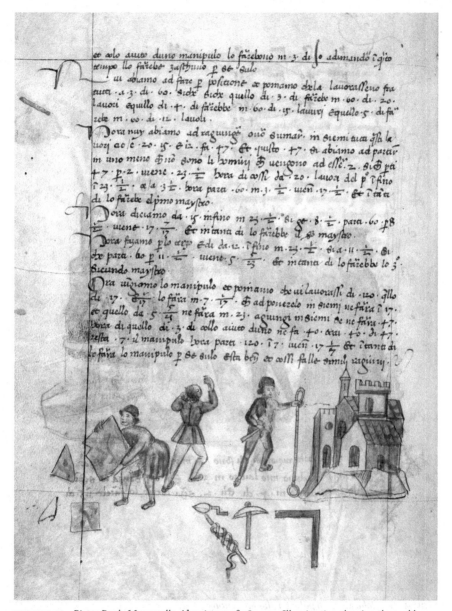

FIGURE 40   *Pietro Paolo Muscarello*, Algorismus, *f. 58* verso. *Illumination showing the problem relating the construction of houses and palaces. Pencil, ink, wash on parchment.*

FIGURE 41   *Nola, Mastrilli palace. 15th–18th century.*
PHOTO HISTANTARTSI ARCHIVE.

FIGURE 42 *Nola, Ceva Grimaldi palace in Via Santa Chiara. 15th century.*
PHOTO HISTANTARTSI ARCHIVE.

FIGURE 43 *Nola, Ceva Grimaldi palace in Via Santa Chiara. 15th century.*
PHOTO HISTANTARTSI ARCHIVE.

FIGURE 44   *Aversa, palace in Piazza San Domenico, 17th–18th centuries.*
PHOTO HISTANTARTSI ARCHIVE.

FIGURE 45   *Carinola, Petrucci palace. 15th century.*
PHOTO HISTANTARTSI ARCHIVE.

FIGURE 46   *Aversa, Azzolini palace, 14th–18th centuries.*
PHOTO HISTANTARTSI ARCHIVE.

FIGURE 47   *Aversa, palace in Via Monserrato, 15th century.*
PHOTO HISTANTARTSI ARCHIVE.

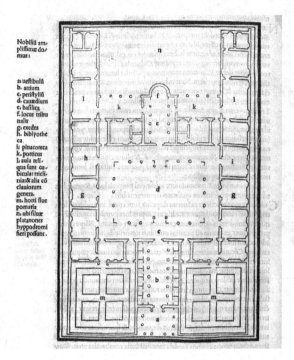

Nobiliū am/
pliſſimæ do/
mus :

a: ueſtibulū
b. atrium
c. periſtyliū
d. cauædium
e. baſilica
f. locus tribu
nalis
g. exedra
h. biblyothe
ca
i: pinacoteca
k. porticus
l. aula reli/
qua ſunt ca/
bicula: tricli/
nia:& alia cō
clauiorum
genera.
m. horti ſiue
pomaria
n. ubi ſiluæ
platanones
hyppodromi
fieri poſſunt .

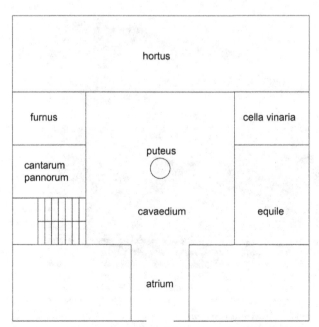

FIGURE 49   *Reconstruction plan of the* domus nolana. *Ground floor
(elaboration Bianca de Divitiis).*

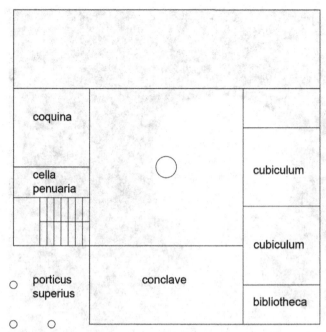

Reconstruction plan of the domus *nolana. First floor*
(*elaboration Bianca de Divitiis*).

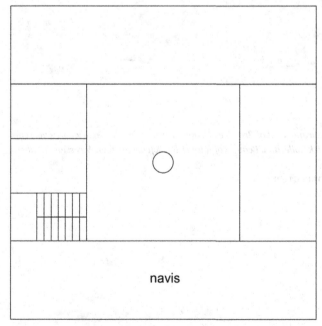

Reconstruction plan of the domus *nolana. Third floor*
(*elaboration Bianca de Divitiis*).

FIGURE 52   *Francesco Laurana, Bust of Beatrice d'Aragona, c. 1475. Marble. 40.6 × 40.3 × 20.3 cm.*
New York, Frick Collection, Bequest of John D. Rockefeller, Jr., 1961. Accession number:
1961.2.86.

FIGURE 53   *Francesco Laurana, Bust of a Lady. c. 1484–96. Marble, H.: 0,44 m.; L.: 0,44 m.;*
             Pr.: 0,24 m. Paris, Louvre.
             © SCALA ARCHIVE.

FIGURE 54
Francesco Laurana (with the
collaboration of Tommaso Malvito),
St. Martha and the Magdalene. Marble,
Marseille, Saint Lazare Chapel.
© BIBLIOTHECA HERTZIANA.

FIGURE 55
Francesco Laurana (with the
collaboration of Tommaso Malvito),
St. Martha and the Magdalene. Marseille,
Saint Lazare Chapel. Marble.
© BIBLIOTHECA HERTZIANA.

FIGURE 56   *Francesco Laurana (with the collaboration of Tommaso Malvito), Tabernacle.*
Marseille, Saint Lazare Chapel, marble.
© BIBLIOTHECA HERTZIANA.

FIGURE 57
Gian Cristoforo Romano,
Beatrice d'Este. H.: 0,59 m
Paris, Louvre, M.L. 10.
© WIKI COMMONS.

- **The inkwell described by Ambrogio Leone (1525)**

Ganymede	Lapiths and Centaurs	Hercules and Cacus	Hercules and Nemean Lion

This inkwell may have been a four-sided box with each section containing figurative panels. Nothing is known about the framing elements or about the objects overall structure which was presumably quite elaborate.

- **The inkwell described in the Francesco Foppa Inventory (1586)**

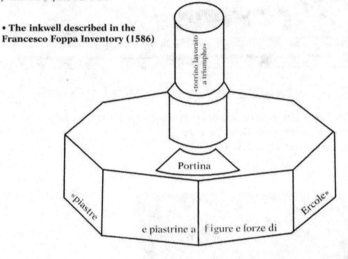

This inkwell consisted of a central tower («torrino») set on a base. Access to the interior of the lower structure was through a «portina» containing a depiction of one of the Labors of Hercules («lavorata de relevo»). The eight-sized base was decorated with «quattro piastre lavorate a figure» and «tre altre piastrine piccole» of the Labors of Hercules. It may or may not be that the «figure» alternated with the «forze». The inventory accounts, however, for only seven of the eight compartments. Furthermore the precise meaning of the distinction between the «piastre» and the «piastrine piccole» remains ambiguous. Given that nothing is known of the overall structure of the «calamaio» into which the figurative elements were placed, the present diagram can be regarded only as a poetic evocation.

FIGURE 58   *Reconstruction of Caradosso's inkwell from Brown, Hickson 1997.*

FIGURE 59
Caradosso, Battle of Centaurs and Lapiths, bronze (cm 4,8 × 4,9), Washington DC, National Gallery of Art, INV. 1957.14.220.
© NATIONAL GALLERY OF ART.

FIGURE 60   *Caradosso, Rape of Ganymede. 8.5 × 8.4 cm Washington, National Gallery of Art), Samuel H. Kress Collection, inv. 1957.14.219.*
© NATIONAL GALLERY OF ART.

FIGURE 61   *Caradosso, Hercules and Cacus. Late 15th century–early 16th century.*
Bronze. Overall dimension (rectangle w/ arched top) 6.1 × 3.7 cm.
Washington, National Gallery of Art, Samuel H. Kress Collection,
1957.14.228.

FIGURE 62 *Medaglia di Nicola Orsini di Pitigliano, Washington DC,*
National Gallery of Art, mm 42,5, Samuel H. Kress Collection,
1957.14.789.a.

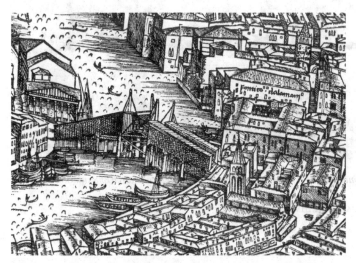

FIGURE 63 *Jacopo de' Barbari,* Venetie MD, *woodcut, 1500. Detail of the*
area of Rialto with the old wooden bridge, the 12th century
Fondaco dei Tedeschi (fontico dalamanj) *and the medieval*
church of St. Bartolomeo (s. / bortolo/mio).

FIGURE 64 *Antonio Minello, Monument of Niccolò Orsini. Venice, Church of Santi Giovanni e Paolo.*
© SCALA ARCHIVE.

FIGURE 65   *Brescia, Museo di Santa Giulia. Monument of Niccolò Orsini.*
© ARCHIVIO FOTOGRAFICO CIVICI MUSEI D'ARTE E STORIA DI
BRESCIA—FOTOSTUDIO RAPUZZI.

FIGURE 66
Fiano Romano, Church of
Santo Stefano, Monument
of Niccolò Orsini.
© BIBLIOTHECA
HERTZIANA.

FIGURE 67   *Venice, Santi Giovanni e Paolo, Stained glass window (lower part).*
              Girolamo Mocetto.
              © SCALA.

FIGURE 68   *Venice, Santi Giovanni e Paolo, Stained glass window. Signature*
              of Girolamo Mocetto.
              PHOTO FULVIO LENZO.

FIGURE 69   *Arms of the families in Nola. Families Albertini, Alfano, Barone, and Cesarini.*
              Naples, Biblioteca Nazionale, ms. X.A.42 f. 91r.

FIGURE 70 *Arms of the families in Nola. Families Fellecchia, Iodice, Fontanagrossa, Girardo, Monforte, Frezza, Mastrilli, Gioseppo, Notare, Palma, Mazzeo, Ferraro, Cannido, Mauro, Piergioanni, Peraro, and Rega. Naples, Biblioteca Nazionale, ms. X.A.42 f. 92r.*

FIGURE 71   *Arms of the families in Nola. Families Sussulano, Tansillo, Ostuni, Bisantizzo,*
Carduccio, Benedicto, Cimino, Bovio, Larcario, Bagniarda, Scalone, Petrarola,
Pitriello, and Zaccaria. Naples, Biblioteca Nazionale, ms. X.A.42 f. 93r.

Bibliography

A   Editions of Works by Ambrogio Leone

A.1   *De Nola* (with Reprints and Modern Editions)

Leone, *De Nola opusculum distinctum, plenum, clarum, doctum, pulcrum* [sic]*, verum, grave, varium, et utile*, Venice: Giovanni Rosso, 4 September 1514.

Leone, *Ambrosii Leonis Nolani de Nola libri III*, in *Italiae illustratae seu Rerum urbiumque Italicarum scriptores varii, notae melioris nunc primum collecti simulque editi*, Frankfurt am Main: Andreas Cambier, 1600, cols. 875–990.

Leone (ed. Graevius), *Ambrosii Leonis Nolani antiquitatum nec non historiarum urbis ac agri Nolae, ut et de montibus Vesuvio et Abella descriptionis, libri tres*, in *Thesaurus antiquitatum et historiarum Italiae*, ed. by Joannes Georgius Graevius, Tomus 9, pars 4, Leiden: Pierre Vander Aa, 1723.

Leone (ed. Jordanus), Ambrosius Leo, *De agro Nolano*, ed. by Dominicus Jordanus in *Delectus scriptorum rerum Neapolitanarum, qui populorum ac civitatum res antiquas, aliasque vario tempore gestas memoriae prodiderunt*, Naples: Francesco Ricciardi, 1735, 1–118.

Leone (ed. Jordanus), Ambrosius Leo, *De agro Nolano*, ed. by Dominicus Jordanus in *Rarissimorum Scriptorum rerum Neapolitanarum collectio qui populorum ac civitatum res antiquas aliasque vario tempore gestas memoriae prodiderunt*, Naples: Nicolò and Vincenzo Rispoli, 1738, 1–118.

Leone (ed. Barbati), Ambrogio Leone, *Nola: la terra natia. Opera piccola, precisa, completa, chiara, dotta, veritiera, seria, varia ed utile*, translated into Italian by Paolino Barbati, Naples 1934.

Leone (ed. Ruggiero), Ambrogio Leone, *Nola*, ed. and translated into Italian by Andrea Ruggiero, Naples 1997.

A.2   Other Works (First Editions Only)

Leone, *Ambrosii Leonis Nolani Marini filii Castigationum adversus Averroem ad Augustissimum Leonem Decimum Pontificem Maximum liber primus in ea quae super praedicamentis ille dixit Averrois*, Venice: Bernardino and Matteo Vitali, 25 September 1517.

Leone, *In Actuarium Joannem Zachariae filium de urinis liber primus, Ambrosio Leone Nolano Marini filio interprete*, Venice: Bernardino Vitali, 10 November 1519.

Leone, *Ambrosii Leonis Nolani divini philosophi Novum opus quaestionum seu problematum ut pulcherrimorum ita utilissimorum tum aliis plerisque in rebus cognoscendis tum maxime in philosophia et medicina scientia*, Venice: Bernardino and Matteo Vitali, 28 August 1523.

Leone, *Lector diligentissime haec est tertia pars magni operis castigationum quas fecit Ambrosius Leo Nolanus philosophus praestantissimus in ea quae scripsit Averroes super Aristotele*, Venice: Bernardino Viani, 20 October 1524.

Leone, *Eximii doctoris Ambrosii Leonis Nolani De nobilitate rerum dialogus. Eiusdem ex Aristotele translatum opus de virtutibus*, Venice: Melchiorre Sessa and Pietro Ravani, September 1525.

Leone, *Anitii Manlii Severini Boethi, ... Dialectica, nunc denique post tot impressiones, auxilio manu scriptorum codicum, in pristinum candorem, a Martiano Rota restituta: ac libris, titulis, ordine, uitaque autoris decorata.... In hac autem novissima impressione si quid forte latuerat (quod in tanta colluvie oportuit) prorsus repositum est ..., Index postremo,... Quibus adiecta est Disputatio Alexandri Aphrodisii de diffinitione, cum annotationibus Ambrosii Leonis in eandem, numquam alias impressam*, Venice: Giovanni Griffio, 1549.

B   General Bibliography

Accame, Dell'Oro, *I Mirabilia urbis Romae*, ed. by Maria Accame, Emy Dell'Oro, Tivoli 2004.

Accame, Maria, "Pomponio Leto, Giulio", in *Dizionario Biografico degli Italiani*, vol. 84, Rome 2015, 711–716.

Adinolfi, Raffaele, *Il tempio di Augusto a Pozzuoli. Raccolta delle fonti iconografiche, epigrafiche, storiche*, Pozzuoli 1968.

Agosti, Giovanni, *Bambaia e il classicismo lombardo*, Turin 1990.

Agosti, Giovanni, "Su Mantegna. 3", *Prospettiva*, 73–74 (1994), 131–143.

Agosti, Giovanni, "Scrittori che parlano di artisti, tra Quattro e Cinquecento in Lombardia", in *Quattro pezzi lombardi (per Maria Teresa Binaghi)*, Brescia 1998, 39–93.

Agosti, Giovanni, *Su Mantegna I*, Milan 2005.

Alberti, Leandro, *Descrittione di tutta Italia ...*, Bologna: Anselmo Giaccarelli, January 1550.

Alberti, Leon Battista, *Descriptio urbis Romae*, ed. by Jean-Yves Boriaud, Francesco Furlan, Florence 2005.

Alberti, Leon Battista, *De re aedificatoria / L'architettura*, ed. by Giovanni Orlandi, Milan 1966.

Albertini, Francesco, *Opusculum de mirabilibus novae et veteris Urbis Romae*, Rome: Giacomo Mazzocchi, 4 February 1510.

Albonico, Simone, "Appunti su Ludovico il Moro e le lettere", in *Ludovicus Dux*, ed. by Luisa Albonico, Vigevano 1995, 66–91.

Aldimari, Biagio, *Memorie historiche di diverse famiglie nobili*, Naples: Giacomo Raillard, 1691.

Aldo Manuzio: il rinascimento di Venezia, catalogue of the exhibition (Venice, 19 March-19 June 2016) ed. Guido Beltramini, Davide Gasparotto, Venice 2016.

Almagià, Roberto, "Studi di Cartografia napoletana", *Archivio Storico per le Province Napoletane*, 37 (1912), 564–592; 38 (1913), 3–35, 318–348, 409–440, 639–654.

Almagià, Roberto, "Un'antica carta topografica del territorio veronese", *Accademia Nazionale dei Lincei, Rendiconti (Classe di scienze morali, storiche e filologiche)*, s. 5, 32 (1925), 63–83.

Amato, Lorenzo, "Francesco Albertini e l'*Opusculum de mirabilibus urbis Romae*: modelli e fonti", in *Acta Conventus Neo-Latini Upsaliensis. Proceedings of the Fourteenth International Congress of Neo-Latin Studies (Uppsala 2009)*, ed. by Astrid Steiner-Weber, 2 vols., Leiden 2012, 167–176.

Amirante, Giosi, *Aversa. Dalle origini al Settecento*, Naples 1998.

Ammirati, Luigi, *Ambrogio Leone nolano*, Marigliano 1983.

Angelillo, Luigi, *La cattedrale di Nola nella sua storia*, Naples 1909.

Antonini, Rosalba, Giuseppe Mollo, Antonia Solpietro, *Il Cippus Abellanus e le Epigrafi latine*, Nola 1997.

Appadurai, Arjun, *Modernity at Large: Cultural Dimensions of Globalization*, Minneapolis 1996.

Avella, Leonardo, *Nola ipotesi, 2. Rivisitazione e nuove ipotesi di alcune mappe riguardanti la cinta muraria della città preromana e romana, parte del territorio adiacente precedentemente prospettato ed altre della città medioevale con alcune considerazioni su quella rinascimentale magistralmente descritta da Ambrogio Leone*, Naples 1990.

Avella, Leonardo, *Fototeca nolana: archivio d'immagini dei monumenti e delle opere d'arte della città e dell'agro*, 12 vols., Naples 1996–1999.

Bakhtin, Mikhail, *Rabelais and His World*, transl. by H. Iswolsky, Cambridge (MA) – London 1968.

Ballacchino, Katia, "Embodying Devotion, Embodying Passion. The Italian Tradition of The *Festa dei Gigli* in Nola", in *Encounters of Body and Soul in Contemporary Religious Practices. Anthropological Reflections*, ed. by Anna Fedele, Ruy Llera Blanes, New York – Oxford 2011, 43–66.

Ballacchino, Katia, "Mettere in valore una comunità in 'questua'. Patrimonio immateriale e cortocircuiti glocali", in *Alla fiera delle identità: patrimoni, turismo, mercati*, ed. by Letizia Bindi, *Voci*, 10 (2013), 21–35.

Barbaro, Ermolao, *Paraphrasis Themistii in Posteriora Aristotelis, in Physica …, interprete Hermolao Barbaro*, Treviso: Bartholomaeus Confalonerius and Morellus Gerardinus, 1481.

Barkan, Leonard, "Rome", in *The Classical Tradition*, ed. by Anthony Grafton, Glenn W. Most, Salvatore Settis, Cambridge (MA) 2010, 839–850.

Barthélemy, L., "La chapelle Saint-Lazare à l'ancienne cathédrale de Marseille", *Bulletin monumental*, s. 5, 12 (1884), 626–637.

Basile, Tania, *Antonio Tebaldeo. Rime della vulgata*, 2 tomi [vol. II of *Antonio Tebaldeo. Rime*, edited by Tania Basile, Jean-Jacques Marchand], Modena 1992.

Basile, Tania, Jean-Jacques Marchand, *Antonio Tebaldeo. Rime. Introduzione* [vol. I of *Antonio Tebaldeo. Rime*, edited by Tania Basile, Jean-Jacques Marchand], Modena 1989.

Beloch, Karl Julius, *Campanien. Geschichte und Topographie des antiken Neapel und seiner Umgebung*, Breslau 1890.

Beltramini, Guido, Howard Burns, Davide Gasparotto (eds.), *Pietro Bembo e le arti*, Venice 2013.

Beltramini, Guido, Howard Burns, Adolfo Tura (eds.), *Pietro Bembo e l'invenzione del Rinascimento*, Catalogue of the exhibition (Padova, Palazzo del Monte di Pietà, 2 febbraio-19 maggio 2013), Venice 2013.

Benzoni, Gino, "Venezia, 11 agosto 1508: mille orecchie per Luca Pacioli", *Studi Veneziani*, n.s., 69 (2014), 59–326.

Bernard-Pradelle, Laurence, "L'influence de la Seconde Sophistique sur la *Laudatio Florentinae Urbis* de Leonardo Bruni", *Rhetorica*, 18 (2000), 355–387.

Bertolini, Lucia, "Primo (e provvisorio) elenco di copisti e possessori di codici contenenti opere albertiane", in *Leon Battista Alberti. La biblioteca di un umanista*, ed. by Roberto Cardini, Florence 2005, 57–82.

Bertolini, Lucia, "Alberti e le *humanae litterae*", in *Leon Battista Alberti e l'architettura, catalogue of the exhibition (Mantova, 16 september 2006–14 January 2007)*, ed. by Massimo Bulgarelli et alii, Cinisello Balsamo 2006, 20–31.

Bertoni, Giulio, *L'Orlando furioso e la Rinascenza a Ferrara*, Modena 1919.

Bertoni, Giulio, *Guarino da Verona fra letterati e cortigiani a Ferrara (1429–1460)*, Genève 1921.

Bianca, Concetta, "Marchese, Francesco Elio", in *Dizionario Biografico degli Italiani*, vol. 69, Rome 2007, 564–566.

Bianca, Concetta, "Da Firenze a Roma: Francesco Albertini", *Letteratura & Arte*, 9 (2011), 59–70.

Bianco, Monica, Elena Strada (eds.), *'I più vaghi e i più soavi fiori'. Studi sulle antologie di lirica del Cinquecento*, Alessandria 2001.

Biermann, Hartmut, "Eine Villenbeschreibung des Sannazaro und L. B. Albertis *De re aedificatoria*: ein Nachtrag zu *Der runde Hof*", *Mitteilungen des Kunsthistorischen Institutes in Florenz*, 34 (1990), 421–423.

Biffi, Giovanni, *Epistulae et carmina*, Milan: Antonius Zarotus, 10 October 1493.

Biffi, Marco, "Osservazioni sulla lingua di Francesco di Giorgio Martini: la traduzione autografa di Vitruvio", *Studi di gramamtica italiana*, 17 (1998), 37–116.

Biffi, Marco, "La lessicografia storica dell'architettura: i casi di Loggia, Loggiato, Portico, e Porticato", in *Logge e/y lonjas: i luoghi del commercio nella storia della città*, ed. by Giancarlo Cataldi, Roberto Corona, Florence 2002, 59–70.

Biffi, Marco, "La terminologia tecnica dell'Alberti tra latino e volgare", in *Alberti e la cultura del Quattrocento*, ed. by Roberto Cardini, Mariangela Regoliosi, Florence 2007, 655–682.

Biffi, Marco, "Fabio Calvo e Vitruvio: traduzione e traslitterazione", in *Saggi di letteratura architettonica da Vitruvio a Winckelmann*, III, ed. by Howard Burns, Francesco Paolo Di Teodoro, Giorgio Bacci, Florence 2010, 50–51.

Biondo Flavio, *Roma instaurata* [Rome: Printer of Statius, 1471].

Biondo Flavio, *Roma triumphans* [Mantua, or Brescia, about 1473].

Biondo Flavio, *Italia illustrata*, Rome: Johannes Philippus de Lignamine, 1474.

Biondo Flavio, *Roma instaurata, De origine et gestis Venetorum, Italia illustrata*, Verona: Bonino Bonini, 1481–1482.

Biondo Flavio, *De Roma instaurata libri tres ad Eugenium IIII Pontificem Maximum, ... de Italia illustrata opus, ... de gestis Venetorum ad Franciscum Foscari serenissimum ducem inclytumque senatum caetorosque Venetae Reipublicae patritios*, Venice: Gregorio de Gregori, 1 July 1510.

Biondo Flavio, *De Roma triumphante....*, Basel: Hieronymus Froben and Nikolaus Episcopius, 1559.

Biondo Flavio, *Biondo Flavios' Italia Illustrata. Text, translation and commentary, 1. Northern Italy; 2. Central and southern Italy*, ed. by Catherine J. Castner, New York 2005–2010.

Biondo Flavio, *Italia illustrata*, 3 vols., ed. by Paolo Pontari, Rome 2011–2017.

Biondo Flavio, *Roma instaurata. Rome restaurée*, 2 vols., ed. by Anne Raffarin-Dupuis, Paris 2005–2012.

Biondo Flavio, *Italy Illuminated*, 2 vols., ed. and translated into English by Jeffrey A. White, London 2005–2016.

Blessich, Aldo, *La geografia alla corte aragonese di Napoli*, Rome 1897.

Bober, Phyllis Pray, "The *Coryciana* and the Nymph Corycia", *Journal of the Warburg and Courtauld Institutes*, 40 (1977), 223–239.

Bock, Nicolas, *Kunst am Hofe der Anjou-Durazzo. Der Bildhauer Antonio Baboccio (1351– ca. 1423)*, München-Berlin 2001.

von Bode, Wilhelm, "Desiderio da Settignano und Francesco Laurana: zwei italienische Frauenbüsten des Quattrocento im Berliner Museum", *Jahrbuch der Preußischen Kunstsammlungen*, 9 (1888), 209–227.

von Bode, Wilhelm, ["Funde"], *Kunstchronik*, 15 (1904), 269.

Bologna, Ferdinando, "Qualche osservazione sulla lettera di Pietro Summonte a Marcantonio Michiel", in *Libri per vedere. Le guide storico-artistiche della città di Napoli: fonti, testimonianze del gusto, immagini di una città*, ed. by Francesca Amirante, Naples 1995, 181–193.

Bolzoni, Lina, *Poesia e ritratto nel Rinascimento*, ed. by Federica Pich, Bari 2008.

Bolzoni, Lina, *Il cuore di cristallo. Ragionamenti d'amore, poesia e ritratto nel Rinascimento*, Turin 2010.

Bonghi Jovino, Maria, Magda Niro, "Nola", in *Bibliografia topografica della colonizzazione greca in Italia e nelle isole tirreniche*, ed. by Giuseppe Nenci et alii, vol. XII, Pisa-Rome 1993, 373–384.

Bonvesin da la Riva, *Le meraviglie di Milano*, ed. by Paolo Chiesa, Milan 2009.

Borrelli, Carlo, *Vindex Neapolitanae nobilitatis Caroli Borrelli Animadversion in Francisci Elii Marchesii Liber de Neapolitanis familiis*, Naples: Egidio Longo, 1653.

Borsi, Stefano, *Leon Battista Alberti e Napoli*, Florence 2006.

Boskovits, Miklós, *The Origins of Florentine Painting: 1100–1270*, Florence 1993.

Bracciolini, Poggio, *Operum partis primae contenta ... partis secundae contenta*, Argentorati: Johann Knobloch, Johann Schott, 1513.

Bracciolini, Poggio, *Les ruines de Rome. De varietate fortunae livre I*, ed. by Jean-Yves Boriaud; intr. and notes by Filippo Coarelli, Jean-Yves Boriaud, Paris 1999.

Bracciolini, Poggio, *De vera nobilitate*, ed. by Davide Canfora, Rome 2002.

Bresc-Bautier (ed.), *Les Sculptures européennes du Musée du Louvre. Moyen Âge, Renaissance et Temps modernes: Byzance, Espagne, îles Britanniques, Italie, anciens Pays Bas et Belgique, pays germaniques et de l'Europe de l'Est, pays scandinaves, antiques restaurées et copies d'antiques*, ed. by Geneviève Bresc-Bautier, Paris 2006.

Brooke, Rosalind, *The image of Saint Francis. Responses to Sainthood in the Thirteenth Century*, Cambridge 2006.

Brown, Clifford M., Sally Hickson, "Caradosso Foppa (ca. 1452–1526/27)", *Arte Lombarda*, 119 (1997), 9–39.

Brown, Clifford M., Anna Maria Lorenzoni, "Caradosso Foppa and the Roman Mint", *Artibus et historiae*, 22 (2001), 41–44.

Brown, Clifford M., Anna Maria Lorenzoni, "Gian Cristoforo Romano e Mantova. La corrispodenza dell'Archivio Gonzaga", *Civiltà mantovana*, 124 (2007), 61–107.

Bruni, Leonardo, *Laudatio Florentine urbis*, ed. by Stefano U. Baldassarri, Florence 2000.

Bruscalupi, Giuseppe, *Vita di Niccolò III Orsini, conte di Pitigliano*, Siena 1872.

Bruscalupi, Giuseppe, *Monografia storica della Contea di Pitigliano*, ed. by. G. C. Fabriziani, Florence 1906.

Bulgarelli, Massimo, "Orafo del Quattrocento (da Leon Battista Alberti?), Cristo libera l'indemoniato, Parigi Musée du Louvre", in *Masaccio e le origini del Rinascimento*, ed. by Luciano Bellosi, Milan 2002, 218–221.

Buonaguro, Carmela, *Documenti per la storia di Nola (secoli XII–XIV)*, Salerno 1997.

Burke, Peter, *Culture and Society in Renaissance Italy 1420–1540*, London 1972.

Burke, Peter, *Popular Culture in Carly Modern Europe*, Farnham-Burlington 2009.

Burns, Howard, "Pirro Ligorio's Reconstruction of Ancient Rome", in *Pirro Ligorio, Artist and Antiquarian*, ed. by Robert W. Gaston, Milan 1988, 19–92.

Burns, Howard, "Alberti", in *Storia dell'architettura italiana. Il Quattrocento*, ed. by Francesco Paolo Fiore, Milan 1998, 114–165.

Caciorgna, Marilena, Cecilia Filippini, Roberto Guerrini (eds.), *Biografia Dipinta. Plutarco e l'arte del Rinascimento 1400–1550*, La Spezia 2001.

Caianiello, Maria Claudia, "La cinta fortificata di Nola tra Quattrocento e Cinquecento", *Castellum*, 45 (2003), 27–50.

Calabi, Donatella, "Le due piazze di San Marco e di Rialto", *Annali di architettura*, 4–5 (1992–1993), 190–201.

Calepino, Ambrogio, *Ad Librum mos est putidus ...*, Reggio Emilia: Dionigi Bertocchi, 1502.

Calvo, Marco Fabio, *Antiquae urbis Romae cum regionibus simulachrum*, Rome: Ludovico Vicentino, April 1527.

Camodeca, Giuseppe, "I pagi di Nola", in *Modalità insediative e strutture agrarie nell'Italia meridionale in età romana*, ed. by Elio Lo Cascio, Alfredina Storchi Marino, Bari 2001, 413–433.

Campanelli, Maurizio, "Monuments and Histories: Ideas and Images of Antiquity in Some Descriptions of Rome", in *Rome Across Time and Space: Cultural Transmission and the Exchange of Ideas c. 500–1400*, ed. by Claudia Bolgia, Rosamond McKitterick, John Osborne, Cambridge 2011, 35–51.

Campano, Johannes Antonius, *Epistolae et poemata ...*, ed. by Joannes Burchardus Menckenius, Lepzig: Joannes Fridericum Gleditsch, 1707.

Campone, Maria Carolina, "Apporti teologici paoliniani all'evoluzione dell'arte liturgica dei primi secoli: la croce gemmata della cattedrale di Nola", *Rendiconti della Accademia di Archeologia, Lettere e Belle Arti*, n.s., 69, 2000, 13–18.

Campone, Maria Carolina, "L'alfa e l'omega: contaminazioni tra Oriente e Occidente. La croce gemmata nel succorpo della cattedrale di Nola", *Studi sull'oriente cristiano*, 13, 2009, 1, 33–48.

Camporesi, Piero, *Il paese della fame*, Bologna 1985.

Cantabene, Giulia, "Episodi di iconografia nolana: dalle vedute del Cinquecento alla Scuola di Posillipo", in *Iconografia delle città in Campania. Napoli e i centri della provincia*, ed. by Cesare De Seta, Alfredo Buccaro, Naples 2006, 267–276.

Cantatore, Flavia, "Piante e vedute di Roma", in *La Roma di Leon Battista Alberti. Umanisti, architetti e artisti alla scoperta dell'antico nella città del Quattrocento*, catalogue of the exhibition (Roma, Musei Capitolini, 24.6.-16.10.2005), ed. by Francesco Paolo Fiore, Milan 2005, 166–169.

Capaccio, Giulio Cesare, *Illustrium mulierum et illustrium litteris virorum elogia*, Naples: Iacobus Carlinus and Constantinus Vitale, 1608.

Capaldi, Carmela, "Rilievi in calcare dall'anfiteatro di Nola", *Rendiconti della Accademia di archeologia lettere e belle arti*, 73 (2004–2005), 439–468.

Capaldi, Carmela, 'Severo more Doricorum' Espressioni del linguaggio figurativo augusteo in fregi dorici della Campania, Pozzuoli 2005.

Caracciolo, Tristano, Nobilitatis Neapolitanae defensio, ed. by Giuseppe Paladino, in Rerum Italicarum Scriptores, XXII.1, fasc.1–2, Bologna 1934–1935, 139–148.

Carillo, Saverio, "La ricostruzione del Duomo di Nola: 1861–1909", in Tutela e restauro dei monumenti in Campania 1860–1900, ed. by Giuseppe Fiengo, Naples 1993, 355–379.

Carillo, Saverio, "Alcune osservazioni sulla pianta della città allegata al De Nola di Ambrogio Leone", in Nola e il suo territorio: dalla fine del Medio Evo al 17. secolo, momenti di storia culturale e artistica: atti del 2. Corso di formazione per docenti in servizio Didattica e territorio, 10 febbraio-28 maggio 1994, ed. by Tobia Toscano, Nola 1996, 25–43.

Caro Baroja, Julio, El Carnaval. Analisis historico-cultural, Madrid 1979.

Carpo, Mario, "Descriptio urbis Romae. 'Ekfrasis' geografica e cultura visuale all'alba della rivoluzione tipografica", Albertiana, 1 (1998), 121–142.

Carpo, Mario, "Drawing with Numbers: Geometry and Numeracy in Early Modern Architectural Design", Journal of the Society of Architectural Historians, 62 (2003), 448–469.

Cassola, Filippo, "Problemi di storia neapolitana", in Neapolis. Atti del venticinquesimo convegno di studi sulla Magna Grecia (Taranto, 3–7 ottobre 1985), Naples 1986, 37–81.

Castoldi, Massimo, "Per il Beatricium", Quaderni di filologia e lingue romanze, s. 3, 4 (1989), 33–59.

Castoldi, Massimo, "Giunta minima al Beatricium: un sonetto di Giovanni Pincaro e sei epigrammi di Lancinio Curti", Quaderni di filologia e lingue romanze, s. 3, 7 (1992), 49–58.

Catone, Angelo, Il Liber de epidemiis, ed. and translated into Italian by Guido Baldi, Enrico Borlone, Marcello Segré, Roma 1968.

de Celano, Thomas, Vita Prima sancti Francisci, ed. by Stanislao da Campagnola, in Fontes Franciscani, ed. by Enrico Menestò et alii, Assisi 1995.

Cellini, Benvenuto, Opere, ed. by Orazio Bacci, Florence 1901.

Cellini, Benvenuto, I trattati dell'oreficeria e della scultura, ed. by Gaetano Milanesi, Florence 1857.

Censorinus, Dio et alii, De die natali liber aureus ... Neruae Traianique et Adriani Caesaris uitae ex Dione in Latinum uersae a Georgio Merula. Item Vesaeui montis conflagratio ex eodem Merula interprete. Cebetis Thebani tabula. Plutarchi libellus de differentia inter odium, et inuidiam. Basilii oratio de inuidia. Basilii epistola de uita solitaria [Milan, Giovanni Giacomo da Legnano, 1503].

Ceriana, Matteo, "Ganti, Giovanni Cristoforo", in Dizionario Biografico degli Italiani, vol. 52, Rome 1999, 203–211.

Cesarano, Mario, "Il disegno dell'anfiteatro di Nola in alcune pergamene aragonesi di Napoli", *Symbolae Antiquariae*, 4 (2011), 49–82.

Chambers, David S., "The Renaissance Cardinalate from Paolo Cortesi's *De cardinalatu* to the Present", in *The Possessions of a Cardinal*, ed. by Mary Hollingsworth, Carol M. Richardson, University Park 2009, 17–24.

Christiansen, Weppelmann, *The Renaissance Portrait from Donatello to Bellini*, Catalogue of the exhibition (Berlin, New York, 2011), ed. by Keith Christiansen, Stefan Weppelmann, New York 2011.

Cicognara, Leopoldo, *Storia della scultura dal suo risorgimento in Italia fino al secolo di Canova: per servire di continuazione alle opere di Winckelmann e di D'Agincourt*, 8 vols., Prato 1823–1824.

Cirese, Alberto M., *Dislivelli di cultura ed altri discorsi inattuali*, Rome 1997.

Clarke, Georgia, "The Palazzo Orsini in Nola: a Renaissance relationship with Antiquity", *Apollo*, 144 (1996), 44–50.

Clarke, Georgia, *Roman House, Renaissance Palaces*, Cambridge 2003.

Classen, Carl Joachim, *Die Stadt im Spiegel der* Descriptiones *und* Laudes urbium, Hildesheim-Zürich-New York 1980.

Clemente, Pietro, Maria Luisa Meoni, Massimo Squillacciotti, *Il dibattito sul folklore in Italia*, Milan 1976.

Collareta, Marco, "La miniatura di Simone Martini per il Petrarca descritta da Sabba da Castiglione", *Prospettiva*, 53–54 (1988–1989), 334–337.

Collareta, Marco, "Aspetti del paragone al tempo di Tullio Lombardo", in *Tullio Lombardo. Scultore e architetto nella Venezia del Rinascimento*, Atti del convegno di studi (Venezia, 2006), ed. by Matteo Ceriana, Verona 2006, 183–185.

Collareta, Marco, "Caradosso [Foppa, Cristoforo]", in *The Grove dictionary of art, ad vocem* updated in 2010. URL: doi.org/10.1093/gao/9781884446054. article.T013927.

Colonna, Francesco, *Hypnerotomachia Poliphili ubi humana omnia non nisi somnium esse docet atque obiter plurima scitu sane quam digna commemorat*, Venice: Aldo Manuzio, December 1499.

Colonna, Francesco, *Hypnerotomachia Poliphili* (Venice 1499), ed. by Giovanni Pozzi, Lucia Ciapponi, Padova 1980.

Constable, Giles, *The Reformation of the Twelth Century*, Cambridge 1996.

Conti, Simonetta, "Bernardo Silvano y su obra cartográfica", *Revista de Estudios Colombinos*, 5 (2009), 63–74.

Cook, William Robert, *Images of St. Francis of Assisi in painting, stone and glass from the earliest images to ca. 1320 in Italy: a catalogue*, Florence 1999.

della Corte, Matteo, "Dove morì Augusto?", *Rendiconti dell'Accademia di Archeologia Lettere e Belle Arti*, 13 (1933–1934), 67–93.

Cortesi, Paolo, *De cardinalatu ad Iulium secundum pont. max. prooemium*, San Gimignano ("in Castro Cortesio"): Simone Nardi, 15 November 1510.

Coryciana critice edidit, carminibus extravagantibus auxit, praefatione et annotationibus instruxit Josephus Ijsewijn, Rome 1997.

Cosimi, Enrico, "*Fons Augusteus*: Le mura d'arce di Sarno ed il doppio canale di Palma Campania", *Gradus*, 3 (2008), 23–42.

Crawford, Michael H., "Antoine Morillon, Antiquarian and Medallist", *Journal of the Warburg and Courtald Institutes*, 61 (1998), 93–110.

Croce, Benedetto, "Francesco Elio Marchese e il suo opuscolo sulla nobiltà napoletana", [*La Critica. Rivista di Letteratura, Storia e Filosofia diretta da B. Croce*, 20 (1922), 180–192], in *Uomini e cose della vecchia Italia*, Bari 1927, vol. 1, 26–45.

Croce, Benedetto, "Erasmo e gli umanisti napoletani", [in *Gedenkschrift zum 400. Todestage des Erasmus von Rotterdam*, Basel 1936] in *Aneddoti di varia letteratura*, vol. 1, Bari 1953, 166–178.

Cundari, Cesare, "Il Palazzo Petrucci a Carinola. Un episodio autentico di architettura catalana di-svelato", in *Verso un repertorio dell'architettura catalana. Architettura catalana in Campania. Province di Benevento, Caserta, Napoli*, ed. by Cesare Cundari, Rome 2005, 57–78.

D'Ambrosio, Angelo, Raffaele Giamminelli, *Il Duomo di Pozzuoli. Evoluzione del tempio augusteo in chiesa cristiana: episcopium Sancti Proculi*, Pozzuoli 2000.

D'Ovidio, Stefano, "La Madonna di Piedigrotta tra storia e leggenda", *Rendiconti dell'Accademia di Archeologia, Lettere e Belle Arti di Napoli*, 74 (2006–2007), 47–91.

D'Ovidio, Stefano, "Boccaccio, Virgilio e la Madonna di Piedigrotta", in *Boccaccio angioino. Materiali per la storia culturale di Napoli nel Trecento*, ed. by Giancarlo Alfano, Teresa D'Urso, Alessandra Perriccioli Saggese, Bruxelles 2012, 329–346.

D'Ovidio, Stefano, "The Crypta Neapolitana: Perception of a Roman Tunnel throughout History", in *The Legacy of Antiquity. New Perspectives in the Reception of the Classical World*, ed. by Lenia Kouneni, Cambridge 2013, 8–29.

Dalché, Patrick Gautier, "The Reception of Ptolemy's Geography (End of the Fourtheeth to Beginning of the Sixtheenth Century)", in *The History of Cartography*, vol. 3.1, ed. by David Woodward, Chicago 2007, 285–364.

Damianaki, Chrysa, "'Come se fussi viva e pura'. Ritrattistica cortigiana tra Quattro e Cinquecento", *Bibliothèque d'Humanisme et Renaissance*, 60 (1998), 349–394.

Damianaki, Chrysa, *I busti femminili di Francesco Laurana tra realtà e finzione*, Sommacampagna 2008.

Danzi, Massimo, *La biblioteca del cardinal Pietro Bembo*, Geneve 2005.

Defilippis, Domenico, "Tra Napoli e Venezia: Il *De Nola* di Ambrogio Leone", *Quaderni dell'Istituto Nazionale di Studi sul Rinascimento Meridionale*, 7 (1991), 23–64.

Defilippis, Domenico, "Antonio Galateo, la Puglia e Venezia", in *La Serenissima e il regno. Nel V centenario dell'Arcadia di Iacopo Sannazaro. Atti del Convegno di studi, Bari-Venezia, 4–8 ottobre 2004*, ed. by Davide Canfora, Angela Caracciolo Aricò, Bari 2006, 125–151.

De Franciscis, Alfonso, "Nola (Napoli). Sepolcro romano", *Bollettino d'Archeologia*, 52 (1967), 197.

Della Schiava, Fabio, Marc Laureys, "La *Roma instaurata* di Biondo Flavio", *Aevum*, 83 (2013), 643–665.

De Lellis, Carlo, *Discorsi delle famiglie nobili napoletane, Parte Prima*, Naples: Honofrio Savio, 1654; *Parte Seconda*, Naples: Gio. Francesco Paci, 1663; *Parte Terza*, Naples: heirs of Roncagliolo, 1671.

De Lellis, Carlo, *Discorsi postumi* [...] *di alcune poche nobili famiglie*, ed. by Domenico Conforto, Naples: Antonio Gramignani, 1701.

Del Treppo, Mario, Alfonso Leone, *Amalfi medievale*, Naples 1977.

Delle Donne, Roberto, "*Regis servitium nostra mercatura*. Culture e linguaggi della fiscalità nella Napoli aragonese", in *Linguaggi politici, cerimoniali civici e pratiche della politica a Genova e nel Regno di Napoli nel tardo Medioevo*, ed. by Giovanna Petti Balbi, Giovanni Vitolo, Salerno 2007, 91–150.

Demosthenis Orationes duae et sexaginta. Libanii in eas ipsas orationes argumenta. Vita Demosthenis per Libanium. Eiusdem vita per Plutarchum, Venice: Aldo Manuzio, November 1504 [but 1513].

De Ritis, Vincenzo, *Vocabolario napoletano lessigrafico e storico*, vol. I, Naples 1845.

De Vito, Sabina, "La vetrata dei SS. Giovanni e Paolo, la guerra di Cambrai e gli affreschi veronesi del Mocetto", *Notizie da Palazzo Albani*, 15 (1986), 19–30.

Di Cerbo, Cristiana, "L'insediamento francescano di Santa Chiara di Nola e la devozione a Maria Jacobi. Un'ipotesi di lettura", *Annali dell'Istituto italiano di studi storici*, 23 (2008), 107–222.

Di Cerbo, Cristiana, "La Nola degli Orsini tra XIII e XIV secolo: topografia, sistema difensivo, *castrum* e magnificazione della città", *Archivio Storico per le Province Napoletane*, 331 (2013), 1–28.

Di Cerbo, Cristiana, "La cattedrale di Nola tra Altomedioevo e Tardogotico: nuove ipotesi interpretative", *Annali dell'Istituto Italiano per gli Studi Storici*, 28 (2014–2015), 313–351.

de Divitiis, Bianca, *Architettura e committenza nella Napoli del Quattrocento*, Venice 2007.

de Divitiis, Bianca, "I resoconti di guerra come fonte per la storia dell'architettura", in *La battaglia nel Rinascimento meridionale*, ed. by Giancarlo Abbamonte, Joana Barreto, Teresa D'Urso, Alessandra Perriccioli Saggese, Francesco Senatore, Rome 2011, 321–334.

de Divitiis, Bianca, "Architettura e identità nell'Italia meridionale del Quattrocento: Nola, Capua e Sessa", in *Architettura e Identità locali*, II, ed. by Howard Burns, Mauro Mussolin, Florence 2013, 315–331.

de Divitiis, Bianca, "Giuliano da Sangallo in the Kingdom of Naples: Architecture and Cultural Exchange", *Journal of the Society of Architectural Historians*, 74 (2015a), 152–178.

de Divitiis, Bianca, "Memories from the Subsoil: Discovering Antiquities in Fifteenth-Century Naples and Campania", in *Remembering Parthenope. The Reception of Classical Naples from Antiquity to the Present*, ed. by Jessica Hughes, Claudio Bongiovanni, Oxford 2015b, 189–216.

de Divitiis, Bianca, "Rinascimento meridionale: la Nola di Orso Orsini tra ricerca dell'antico e nuove committenze", *Annali di Architettura*, 28 (2016a), 27–48.

de Divitiis, Bianca, "Humanists and artistic debate in XVth century southern Italy", with an appendix by Lorenzo Miletti: "The chapters on architecture and urbanism in Leone's *De nobilitate rerum*", *Humanistica*, 8, 2016b, 153–179.

de Divitiis, Bianca, "A local sense of the past: spolia, re-use and all'antica building in Southern Italy (1400–1600)", in *Local Antiquities, Local Identities: Art, Literature and Antiquarianism in Europe c 1400–1700*, ed. by Kathleen Christian, Bianca de Divitiis, forthcoming (2018).

Donati, Claudio, *L'idea di nobiltà in Italia: Secoli XIV–XVIII*, Bari 1988.

Donati, Gemma, *L'Orthographia di Giovanni Tortelli*, Messina 2006.

Du Cange, Favre, *Glossarium mediae et infimae latinitatis conditum a Carolo du Fresne domino du Cange [...], editio nova aucta pluribus verbis aliorum scriptorum a Léopolde Favre*, Niort 1883–1887.

Dubois, Charles, *Pouzzoles antique. Histoire et Topographie*, Paris 1907.

Durandus , Gulielmus, *Divinorum Officiorum*, ed. by Anselme Davril O.S.B., Timothy M. Thibodeau, vol. I, Turnholt 1995.

D'Uva, Francesco, *I Gigli di Nola e l'Unesco. Il patrimonio culturale immateriale tra politiche internazionali e realtà territoriali*, Nola 2010.

Ebanista, Carlo, "Il ruolo del santuario martoriate di Cimitile nella trasformazione del tessuto urbano di Nola", in *Le città campane fra tarda antichità e alto medioevo*, ed. by Giovanni Vitolo, Salerno 2005, 313–377.

Ebanista, Carlo, "Tra Nola e Cimitile: alla ricerca della prima cattedrale", *Rassegna Storica Salernitana*, 24 (2007), 25–119.

Egnazio, Giovanni Battista, *Oratio habita in funere clarissimi Imperatoris Nicolai Ursini Nolae Pitilianique Principis*, [Venice, after January 1510].

Egnazio, Giovanni Battista, *In Dioscoridem ab Hermolao Barbaro tralatum annotamenta ...*, Venice: Luigi and Francesco Barbaro, 1 February, 1516.

Egnazio, Giovanni Battista, *De Caesaribus libri III ...*, Venice: heirs of Aldo Manuzio and Andrea Torresano, 1516.

Epistolae diversorum philosophorum, oratorum, rhetorum sex et viginti, Venice: Aldo Manuzio 1499.

Erasmus, *Opus epistolarum Des. Erasmi Roterdami*, tom. 3, 1517–1519, ed. by Percy S. Allen, Oxford 1913.

Euclid, *Opera*, ed. by Luca Pacioli, Venice: Paganino Paganini, 11 June 1509.

Falassi, Alessandro, "Feste, teste, tempeste", in *La festa*, ed. by Alessandro Falassi, Milan 1988, 9–29.

Farnoux, Bernard Combet, "Mercure romain, les *Mercuriales* et l'institution du culte impérial sous le principat augustéen", in *Aufstieg und Niedergang der römischen Welt*, 11, Berlin 1987, 457–501.

Fasoli, Gina, "La coscienza civica nelle *Laudes civitatum*", [in *La coscienza cittadina nei Comuni italiani del Duecento*, Todi 1972, 11–44] in Gina Fasoli, *Scritti di storia medievale*, Bologna 1973, 293–318.

Fattorini, Gabriele, "Francesco di Giorgio e la cultura urbinate nel 'giro' degli Orsini", *Predella*, 30 (2011), December. Online version http://predella.arte.unipi.it

Favole, Adriano, *Resti di umanità. Vita sociale del corpo dopo la morte*, Rome-Bari 2003.

Fera, Vincenzo, "Polemiche filologiche intorno allo Svetonio del Beroaldo", in *The Uses of Greek and Latin: Historical Essays*, ed. by Anna Carlotta Dionisotti, Anthony Grafton, Jill Kraye, London 1988, 71–87.

Ferreri, Luigi, *L'Italia degli umanisti. Marco Musuro*, Turnhout 2014.

Fiengo, Giuseppe, *I Regi Lagni e la bonifica della* Campania Felix *durante il viceregno spagnolo*, Florence 1988.

Fiengo, Giuseppe, *L'acquedotto di Carmignano e lo sviluppo di Napoli in età barocca*, Florence 1990.

Fiengo, Giuseppe, Luigi Guerriero, *Il centro storico di Aversa. Analisi del patrimonio edilizio*, Naples 2002.

Fiengo, Giuseppe, Luigi Guerriero, "La residenza aragonese della Conigliera in Napoli", *Napoli Nobilissima*, s. 6, 1 (2010), 81–102.

Figliuolo, Bruno, *La cultura a Napoli nel secondo Quattrocento*, Udine 1997.

Figliuolo, Bruno, "Pontano, Giovanni", in *Dizionario Biografico degli Italiani*, vol. 84, Rome 2015, 729–740.

Filangieri, Gaetano, *Documenti per la storia le arti e le industrie delle provincie napoletane*, 6 vols., Naples 1883–1891.

Finzi, Claudio, "La polemica sulla nobiltà nell'Italia del Quattrocento", *Cuadernos de Filología Clásica. Estudios Latinos*, 30/2 (2010), 341–380.

Finzi, Roberto, *L'onesto porco. Storia di una diffamazione*, Milan 2014.

Folin, Marco, "Piante di città nell'Italia di antico regime: uno strumento di conoscenza analitico-operativa; introduzione", in *Rappresentare la città: topografie urbane nell'Italia di antico regime*, ed. by Marco Folin, Reggio Emilia 2010a, 9–55.

Folin, Marco, "La Proportionabilis et commensurata designatio urbis Ferrariae di Pellegrino Prisciani (1494–1495)", in *Rappresentare la città: topografie urbane nell'Italia di antico regime*, ed. by Marco Folin, Reggio Emilia 2010b, 99–120.

Fontana, Renzo, "Oltre il medioevo: la lega di Cambrai e l'iconografia della guerra", *Quaderni del Centro Studi Medievali Ponzio di Cluny*, n.s., 1 (2009–2010), 133–195.

Fontana, Vincenzo, "Fra Giovanni Giocondo ingegnere idraulico a Venezia (1506–1508)", in *Giovanni Giocondo umanista, architetto e antiquario*, ed. by Pierre Gros, Pier Nicola Pagliara, Venice 2014, 363–379.

Fontanella, Francesca, "Aspetti di storia della fortuna di Elio Aristide nell'età moderna", in *Elio Aristide e la legittimazione greca dell'impero di Roma*, ed. by Paolo Desideri, Francesca Fontanella, Bologna 2013.

Fossaluzza, Giorgio, "La 'Grande vetrata' di San Zanipolo: problemi aperti e nuove acquisizioni a trent'anni dalla mostra al Museo Correr", *Bollettino dei Musei Civici veneziani*, 7 (2012), 82–83.

de Foville, Jean, "La médaille de Nicolas Orsini, Compte de Pitigliano", *Revue Numismatique*, s. 4, 15 (1911), 450.

Fragmenta et picturae Vergiliana Codicis Vaticani Latini 3225 phototypice expressa, consilio et opera curatorum Bibliothecae Vaticanae, Vatican City 1945.

Friedmann, David, "Talamone, 1306", in *Rappresentare la città: topografie urbane nell'Italia di antico regime*, ed. by Marco Folin, Reggio Emilia 2010a, 57–76.

Friedman, David, "La pianta di Imola di Leonardo, 1502", in *Rappresentare la città: topografie urbane nell'Italia di antico regime*, ed. by Marco Folin, Reggio Emilia 2010b, 121–144. stesso autore, stesso anno.

Frommel, Cristof L., Stefano Ray, Manfredo Tafuri, *Raffaello architetto*, (Milan 1984) Milan 2002.

Frugoni, Chiara, *Francesco: un'altra storia*, Genoa 1988.

Frutaz, Amato Pietro, *Le piante di Roma*, Rome 1962.

Fubini, Riccardo, "Biondo Flavio", in *Dizionario Biografico degli Italiani*, vol. 10, Rome 1968, 536–559.

Furlan, Francesco, "In margine all'edizione degli *Ex ludis rerum mathematicarum*: ossia osservazioni e note per l'edizione di un testo scientifico e delle sue figure", *Revue d'histoire des sciences*, 59 (2006), 197–217.

Gaborit-Chopin, Danielle, *Lampe or perfume-burner in shape of dome building*, in *The Treasure of San Marco Venice*, Milan 1984, 237–243, entry 33.

Gaignebet, Claude, *Le Carnaval. Essays de mytologie populaire*, Paris 1974.

Galateo, Antonio de Ferrariis, *Liber de situ Iapygiae. Descriptio urbis Callipolis*, Basel: Petrum Pernam, 1558.

Galateo, Antonio De Ferrariis, *La Iapigia* (*Liber de situ Iapigyae*), ed., translated into Italian and annoted by Domenico Defilippis, Galatina 2005.

Galateo, Antonio De Ferrariis, *De nobilitate*, in *Puglia Neo-Latina: Un itinerario del Rinascimento tra autori e testi*, ed. by Francesco Tateo, Mauro de Nichilo, Pietro Sisto, Bari 1994, 107–175.

Ganda, Arnaldo, "La biblioteca latina del poeta milanese Lancinio Curzio (1462–1512)", *La Bibliofilia*, 93 (1991), 221–277.

Gasparotto, Davide, "Antico e Moderno", in *Bonacolsi l'Antico. Uno scultore nella Mantova di Andrea Mantegna e di Isabella d'Este*, Catalogue of the exhibition (Mantova, 2008–2009), Milan 2008, 88–97.

Gasparotto, Davide, "Pietro Bembo, gli amici, la collezione, gli artisti", in *I cardinali della Serenissima. Arte e committenza tra Venezia e Roma (1523–1605)*, ed. by Caterina Furlan, Patrizia Tosini, Milan 2014, 129–159.

Gaskell, Roger, "Printing House and Engraving Shop. A Mysterious Collaboration", *The Book Collector*, 53, (2004), 213–251.

Gaurico, Pomponio, *De sculptura (1504)*, ed. and translated into French by André Chastel, Robert Klein, Genève 1969.

Genesin, Monica, Luana Rizzo (eds.), *Magie, Tarantismus und Vampirismus. Eine interdisziplinäre Annäherung*, Hamburg 2013.

Gentilini, Anna Rosa (ed.), *Sabba da Castiglione 1480–1554. Dalle corti rinascimentali alla Commenda di Faenza. Atti del Convegno, Faenza 19–20 maggio 2000*, Florence 2006.

Giannini, Maria, "Il 'palazzo senatorio' di Paolo Cortesi. L'architettura nel *De cardinalatu*", *Miscellanea storica della Valdelsa*, 108 (2002), 63–82.

Giovio, Paolo, *Scritti d'arte*, ed. by Sonia Maffei, Pisa 1999.

Glorious Horsemen. Equestrian Art in Europe 1500–1800, Catalogue of the exhibition, Springfield (MA), Louisville, 1981–1982 [essays and catalogue: Laura Camins]. Springfield 1981.

Goldschmidt, Rudolf C., *Paulinus' Churches at Nola. Texts, Translations and Commentary*, Amsterdam 1940.

Golzio, Vincenzo, *Raffaello nei documenti, nelle testimonianze dei contemporanei e nella letteratura del suo secolo*, Vatican City 1936.

Grapaldi, Francesco Mario, *De partibus aedium*, [Parma]: Angelo Ugoleto, [1494].

Gros, Pierre, "Les lectures vitruviennes du XVIE siècle et quelques-unes de leurs conséquences à l'âge classique: l'exemple de la domus", in *Architecture et théorie. L'héritage de la Renaissance* (Actes de colloques), published on line 26 October 2015, consulted 27 October 2015. URL: http://inha.revues.org/3436.

Guadagni, Carlo, *Nola Sagra*, [1688] ed. by Tobia R. Toscano, Massalubrense 1991.

Guarini, Raimondo, *Ricerche sull'antica città di Eclano. Seconda edizione corretta, ed accresciuta dall'autore*, Naples 1814.

Haffner, Thomas, *Die Bibliothek des Kardinals Giovanni d'Aragona (1456–1485). Illuminierte Handschriften und Inkunabeln für einen humanistischen Bibliophilen zwischen Neapel und Rom*, Wiesbaden 1997.

Haig Gaisser, Julia, "The Rise and Fall of Goritz's Feasts", *Renaissance Quarterly*, 48 (1995), 41–57.

Hamilton, Riccioni, *Rome Re-Imagined. Twelfth-Century Jews, Christians and Muslims Encounter the Eternal City*, ed. by Louis I. Hamilton, Stefano Riccioni, Leiden 2012.

Heers, Jacques, *Fêtes des fous et carnavals*, Paris 1983.

Hill, George Francis, *A Corpus of Italian Medals of the Renaissance before Cellini, in 2 volumes with 201 plates*, London 1930.

Hill, George Francis, *The Gustave Dreyfus Collection: Renaissance Medals*, Oxford 1931.

Hill, George Francis, Graham Pollard, *Renaissance medals from the Samuel H. Kress Collection at the National Gallery of Art* [*Washington D.C.*] *based on the catalogue of Renaissance medals in the Gustave Dreyfus Collection by G. F. Hill, revised and enlarged by Graham Pollard*, London 1967.

Hind, Artur M., *Early Italian Engraving. A Critical Catalogue with Complete Reproduction of All the Prints Descrived*, London 1938–1948.

Hofer, Philip, "Early Book Illustration in the Intaglio Medium. Part II", *The Print Collector's Quarterly*, 21 (1934), 295–316.

Homerus, *Opera*, [ed. by Demetrius Chalcondylas], Florence: Bernardo and Nerio Nerli, and Demetrius Damilas, 1488.

Homeri Ilias, Venice: Aldus, 1504.

Hyde, John Kenneth, "Medieval Descriptions of Cities", *Bulletin of the John Rylands Library*, 48 (1965–1966), 308–40.

Iacono, Antonietta, "La *Laudatio urbis Neapolis* nell'appendice archeologico-antiquaria del *De bello Neapolitano* di Giovanni Gioviano Pontano", *Bollettino di Studi Latini*, 39 (2009), 562–586.

Imbriani, Eugenio, *Dimenticare. L'oblio come pratica culturale*, Nardò 2004.

Imbriani, Eugenio, "Il sacco pieno", in *La maschera e il corpo*, ed. by Pietro Sisto, Piero Totaro, Bari 2012, 124–134.

Immarco, Rosanna, "Alcune considerazioni sulla morte di Augusto a Nola", in *Scritti di varia umanità in memoria di Benito Iezzi*, ed. by Mario Capasso, Enzo Puglia, Sorrento 1994, 93–103.

Iuliano, Marco, "Cartapecore geografiche: cartografia calabra in età aragonese", in *Storia della Calabria nel Rinascimento. Le arti nella storia*, ed. by Simonetta Valtieri, Rome 2002, 49–68.

Jacazzi, Danila, "La memoria e le immagini del territorio napoletano nelle pergamene aragonesi", in *Architettura nella storia. Scritti in onore di Alfonso Gambardella*, ed. by Gaetana Cantone, Laura Marcucci, Elena Manzo, Milan 2008, 89–98.

Jacks, Philip J., "The *Simulachrum* of Fabio Calvo: A View of Roman Architecture 'all'antica' in 1527", *The Art Bulletin*, 72 (1990), 453–481.

Kelly, Samantha (ed.), *The Cronaca di Partenope. An Introduction to and a Critical Edition of the First Vernacular History of Naples (c. 1350)*, Leiden-Boston 2011.

Kent, Francis W., Caroline Elam, "Piero del Massaio: Painter, Mapmaker and Military Surveyor", *Mitteilungen des Kunsthistorischen Institutes in Florenz*, 57 (2015), 65–89.

Koortbojian, Michael, "A Collection of Inscriptions for Lorenzo De' Medici. Two Dedicatory Letters from Fra Giovanni Giocondo: Introduction, Texts and Translation", *Papers of the British School at Rome*, 70 (2002), 297–317.

Kruft, Hanno-Walter, *Francesco Laurana: ein Bildhauer der Frührenaissance*, München 1995.

Kusch-Arnhold, Britta, "Bambaias Grabstele für Lancino Curzio und einige Bemerkungen zur Ikonographie der 'fama'", in *Docta Manus. Studien zur italienischen Skulptur für Joachim Poeschke*, Münster 2007, vol. I, 213–223.

La Greca, Ferdinando, Vladimiro Valerio, *Paesaggio antico e medioevale nelle mappe aragonesi di Giovanni Pontano. Le terre del Principato Citra*, Acciaroli 2008.

Landau, David, "L'arte dell'incisione a Venezia ai tempi di Manuzio", in *Aldo Manuzio: il rinascimento di Venezia*, catalogue of the exhibition (Venezia, Gallerie dell'Accademia, 19 marzo-19giugno 2016) ed. by Guido Beltramini, Davide Gasparotto, Venice 2016, 107–136.

Landau, David, Peter Parshall, *The Renaisance Print: 1470–1550*, New Haven – London 1994.

Landino, Cristoforo, *De vera nobilitate*, ed. by Maria Teresa Liaci, Florence 1970.

Lanternari, Vittorio, "I riti dell'anno. Campania", in *La festa*, ed. by Alessandro Falassi, Milan 1988, 152–165.

La Rocca, Eugenio, Dora Angelillo, *Nola dalle origini al Medioevo*, Naples 1971.

Lanzarini, Orietta, Roberta Martinis, *'Questo libro fu di Andrea Palladio.' Il codice Destailleur B dell'Ermitage*, Rome 2015.

Lehmann, Tomas, *Paolino da Nola: poeta architetto e committente delle costruzioni*, in *Anchora vitae*, Atti del II Convegno Paoliniano nel XVI Centenario del Ritiro di Paolino a Nola (Nola—Cimitile 18–20 maggio 1995), ed. by Gennaro Luongo, Nola 1998, 93–104.

Lenci, Angiolo, *Il leone, l'aquila e la gatta. Venezia e la Lega di Cambrai. Guerra e fortificazioni della battaglia di Agnadello all'assedio di Padova del 1509*, Padua 2002.

Lenza, Cettina, "Dal modello al rilievo: la villa di Poggioreale in una pianta della collezione di Pierre-Adrien Pâris", *Napoli Nobilissima*, s. 5, 5 (2004), 177–188.

Lenzo, Fulvio, "Aggiornamento", in Anthony Blunt, *Architettura barocca e rococò a Napoli*, Milan 2006, 270–329.

Lenzo, Fulvio, *Architettura e antichità a Napoli dal XV al XVIII secolo. Le colonne del tempio dei Dioscuri e la chiesa di San Paolo Maggiore*, Rome 2011.

Lenzo, Fulvio, *Memoria e identità civica. L'architettura dei seggi nel Regno di Napoli, XIII–XVIII secolo*, Rome 2014.

Lenzo, Fulvio, "Philibert de l'Orme et les architectures antiques et moderne du Royaume de Naples", *Revue de l'Art*, 188 (2015), 41–47.

Lepori, Fernando, "La scuola di Rialto dalla fondazione alla metà del Cinquecento", in *Storia della cultura veneta*, vol. III/2, ed. by Girolamo Arnoldi, Manlio Pastore Secchi, Vicenza 1981, 597–600.

Lettieri, Pierantonio, "Discorso dottissimo del Magnifico MS Pierro Antonio de Lecthiero cittadino et Tabulario Napolitano circa l'anticha pianta et ampliatione dela Città di Nap. et del itinerario del acqua che anticamente flueva et dentro et fora la pred[ett] città per aquedocti mirabili quale secondo per più raggioni ne dimostra, era il Sebbetho celebrato dagli antichi auttori", manuscript ca. 1560 ed. in Lorenzo

Giustiniani, *Dizionario geografico ragionato del Regno di Napoli*, vol. VI, Naples 1803, 382–411.

Lewis, Douglas, "The plaquettes of 'Moderno' and his followers", in *Italian Plaquette*, ed. by Alison Luchs, Washington D.C. 1989, 105–141.

Lewis, Douglas, "Collectors of Renaissance Reliefs: Erasmus of Rotterdam (1466–1536) and Baron Boissen de Monville (1763–1832)", in *Collecting Sculpture in Early Modern Europe*, ed. by Nicolas Penny, Eike D. Schmidt, New Haven-London 2008, 129–143.

Litta, Pompeo, *Famiglie celebri di Italia*, Milan 1819–1883.

Lomazzo, Giovanni Paolo, *Scritti sulle arti*, ed. by Roberto Paolo Ciardi, Florence 1973–1975.

Luciano, Pietro et alii, "L'acquedotto di San Paolino e la problematica della distribuzione delle acque nel territorio", in *Primo Convegno dei gruppi archeologici campani*, Rome 1980, 123–132.

Lücke, Hans-Karl, *Alberti index. De re aedificatoria (Florenz 1485), Index verborum*, Munchen 1975–1979.

Maffei, Raffaele, *Commentariorum rerum urbanarum libri XXXVIII*, Romae: Johann Besicken, 1506.

Maffei, Sonia, "La villa di Poggioreale e la Duchesca di Alfonso II di Aragona in una descrizione di Paolo Giovio. Moduli dell'elogio e tradizione antica", *Annali della Scuola Normale Superiore di Pisa. Classe di Lettere e Filosofia. Quaderni*, s. 4, 1–2 (1996), 161–182.

Maiuri, Amedeo, *I Campi Flegrei. Dal sepolcro di Virgilio all'antro di Cuma*, (Rome 1934), Rome 1958.

Maiuri, Amedeo, "Virgilio e Nola", *Quaderni di Studi Romani*, 4 (1939), 7–9.

Maiuri, Amedeo, "Nola. Città e contado", in Amedeo Maiuri, *Passeggiate campane*, Florence 1957, 259–265.

Maiuri, Amedeo, "Sul *De Nola* di Ambrogio Leone", in *Studi in onore di Riccardo Filangieri*, vol. II, Naples 1959, 261–271.

Malaguzzi Valeri, Francesco, *La corte di Ludovico il Moro. Gli artisti lombardi*, Milan 1917.

Manetti, Antonio, *Vita di Filippo Brunelleschi preceduta da La novella del Grasso*, ed. by Domenico De Robertis, Giuliano Tanturli, Milan 1976.

Manetti, Giannozzo, *De Dignitate et excellentia hominis*, ed. by Elizabeth R. Leonard, Padua 1975.

Manutius, Aldus, *The Greek Classics*, ed. and translated into English by Nigel G. Wilson, Cambridge (MA)—London 2016.

Manzi, Pietro, "Alcuni documenti di cartografia nolana, ovvero Ambrogio Leone e Girolamo Moceto", *L'Universo*, 53 (1973), 811–818.

Manzi, Pietro, "Editori tipografi e librai napoletani a Venezia nel secolo XVI", *La Bibliofilia*, 76, (1974), 35–138.

Marano, Francesco, *La Uglia. Riti di attraversamento del fuoco in Lucania*, Potenza 1997.

Marchand, Jean-Jacques, *Rime estravaganti*, 2 tomi [vol. III of *Antonio Tebaldeo. Rime*, edited by Tania Basile and Jean-Jacques Marchand], Modena 1992.

Markham Schulz, Anne, "Four New Works by Antonio Minello", *Mitteilungen des Kunsthistorischen Institutes in Florenz*, 31 (1987), 291–326.

de Marinis, Tammaro, *La biblioteca napoletana dei re d'Aragona*, I, Milan 1952.

Martines, Giangiacomo, "Fonti iconografiche antiche per la ricostruzione rinascimentale della città vitruviana", *Ricerche di storia dell'arte*, 1–2 (1976a), 277–285.

Martines, Giangiacomo, "*Gromatici veteres*", *Ricerche di storia dell'arte*, 3 (1976b), 3–23.

Martini, Giuseppe, "Nola nel secondo Quattrocento", *Nuova Rivista Storica*, 65 (1981), 293–324.

de Martino, Ernesto, *The Land of Remorse: A Study of Southern Italian Tarantism*, (Milan 1961), translated into English by Dorothy Zinn, London 2005.

de Martino, Ernesto, *La terra del rimorso. Contributo a una storia religiosa del Sud*, (Milan 1961(Milan 2008).

de Martino, Ernesto, *Morte e pianto rituale. Dal lamento funebre antico al pianto di Maria*, Turin 1983.

Maurello, Giovanni, *Canzoni in morte di Enrico d'Aragona*, Cosenza: Octavianus Salamonius, after 11 May 1478.

Mazzocco, Angelo, "The antiquarianism of Francesco Petrarca", *The Journal of Medieval and Renaissance Studies*, 7, 1977, 203–224.

Mazzocco, Angelo, "Biondo Flavio and the Antiquarian Tradition", in *Acta conventus neo-latini Bononiensis. Proceedings of the Fourth International Congress of Neo-Latin Studies. Bologna, 26 August to 1 September 1979*, ed. by Richard J. Schoeck, Binghamton (NY) 1985, 122–136.

Mazzocco, Angelo, Marc Laureys (eds.), *A New Sense of the Past: The Scholarship of Biondo Flavio (1392–1463)*, Leuven 2015.

Melfi, Eduardo, "Curti, Lancino", in *Dizionario Biografico degli Italiani*, vol. 31, Rome 1985, 487–488.

Menander Rhetor, ed. with transl. and comm. by Donald A. Russell, Nigel G. Wilson, Oxford 1981.

Middeldorf, Ulrich, Oswald Goetz, *Medals and Plaquettes from the Sigmund Morgenroth Collection*, Chicago 1944.

Miglio, Massimo, *Pellegrinaggi a Roma. Il codice di Einsiedeln, L'itinerario di Sigerico, L'Itinerario Malmesburiense, Le meraviglie di Roma, Racconto delle meraviglie della città di Roma*, Rome 1999.

Miletti, Lorenzo, "Sulla fortuna di Livio nel Cinquecento. Le *domus* dei nobili capuani nella veduta di *Capua vetus* di Cesare Costa", *Bollettino di Studi Latini*, 44 (2014), 107–126.

Miletti, Lorenzo, "Peto (Poetus, Paetus), Francesco (Franciscus)", in *Dizionario Biografico degli Italiani*, vol. 82, Rome 2015, 665–667.

Miletti, Lorenzo, "Virgil and the water of Nola. The Renaissance debate between Pontano, Sannazaro, Ambrogio Leone, and Erasmus on *Georgics* 2.224–5", *Philologus*, 160 (2016a), 337–359.

Miletti, Lorenzo, "Nola città augustea nel Rinascimento meridionale: intorno al *De Nola* di Ambrogio Leone", *Maia*, 68.2 (2016b), 594–605.

Miletti, Lorenzo, "Il *De bello Neapolitano* di Pontano e le città del Regno di Napoli", in *Linguaggi e ideologie del Rinascimento monarchico aragonese (1442–1503): forme della legittimazione e sistemi di governo*, ed. by Fulvio Delle Donne, Antonietta Iacono, Naples 2018, 63–79.

Miletti, Lorenzo, "*Tacitis regnavit Amyclis*. Francesco Peto da Fondi su Virgilio *Aeneis*, X 563–4", in *Le modèle et les originaux. L'exemplum virgilien à Naples entre dynastie aragonaise et vice-royaume espagnol*, ed. by Marc Deramaix, Giuseppe Germano, Paris, forthcoming.

Miltenov, Petrana, "L'agro nolano", in *I centri storici della provincia di Napoli. Struttura, forma e identità urbana*, ed. by Cesare De Seta, Alfredo Buccaro, Naples 2009, 315–34.

Modesti, Paola, "Quasi come in un dipinto: la città e l'architettura nel 'De situ urbis Venetae' di Marcantonio Sabellico", *Arte Veneta*, 66 (2009), 17–35.

Modesti, Paola, *Le delizie ritrovate. Poggioreale e la villa del Rinascimento nella Napoli aragonese*, Florence 2014.

Moffitt, John F., "Medieval Mappaemundi and Ptolemy's Chorographia", *Gesta*, 32 (1993), 59–68.

Mollo, Giuseppe, "Note sull'Architettura di palazzo Albertini in Nola", in *Nola e il suo territorio: dalla fine del Medio Evo al 17. secolo, momenti di storia culturale e artistica: atti del 2. Corso di formazione per docenti in servizio Didattica e territorio, 10 febbraio–28 maggio 1994*, ed. by Tobia Toscano, Nola 1996, 53–62.

Molteni, Elisabetta, "Venezia, fondaco dei Tedeschi. Le ricostruzioni di un edificio nel cuore della città", in Francesco Dal Co, Elisabetta Molteni, *Il Fondaco dei Tedeschi, Venezia, OMA. Il restauro e il riuso di un monumento veneziano*, Milan 2016, 12–79.

Mommsen, Theodor, *Corpus Inscriptionum Latinarum. Vol. X: Inscriptiones Bruttiorum, Lucaniae, Campaniae....*, Berolini 1883.

Montanari, Massimo, *Il formaggio con le pere. La storia di un proverbio*, Rome-Bari 2008.

de Montera, Pierre, "La Beatrice d'Ambroise Leone de Nola: ce qui reste d'un 'Beatricium' consacré à sa gloire", in *Mélanges de philologie, d'histoire et de littérature offerts à Henri Hauvette*, Paris 1934, 191–210.

Morelli, Iacopo, *Notizia d'opere di disegno nella prima metà del secolo XVI esistenti in Padova, Cremona, Milano, Pavia, Bergamo, Crema e Venezia, scritta da un anonimo di quel tempo, pubblicata e illustrata da D. Iacopo Morelli custode della Regia Biblioteca di San Marco di Venezia*, Bassano 1800.

Mortimer, Ruth, *Catalogue of Books and Manuscripts / Harvard College Library, Department of printing and graphic arts*, ed. by Ruth Mortimer, Part 2, *Italian 16. century books*, Cambridge (MA) 1974.

Muntz, Eugène, "L'orfèvrerie romaine de la Renaissance", *Gazette des Beaux-Arts*, 27 (1883), 411–424.

Muratori, Lodovico Antonio, *Della perfetta poesia italiana*, tomo secondo, Modena: Bartolomeo Soliani, 1706.

Muscarello, Pietro Paolo, *Algorismus. Trattato di aritmetica pratica e mercantile del secolo XV*, ed. by Giorgio Chiarini, Verona 1972.

Nagel, Alexander, *Michelangelo and the Reform of Art*, Cambridge 2000.

Nencioni, Giovanni, "Sulla formazione di un lessico nazionale dell'architettura", *Bollettino d'informazioni del Centro di Ricerche Informatiche per i Beni Culturali della Scuola Normale Superiore di Pisa*, 5 (1995), 7–33.

Nenna, Giovanni Battista, *Il Nennio, nel quale si ragiona di nobiltà*, Venice: Giovanni Andrea and Florio Valvassori, 1542.

Nicolini, Fausto, *L'arte napoletana del Rinascimento e la lettera di Pietro Summonte a Marcantonio Michiel*, Naples 1925.

de Nolhac, Pierre, "Le Virgile du Vaticane et ses peintures", *Notice et extraits des Manuscripts de la bibliothèque Nationale et autres bibliothèque*, 35 (1897), 683–791.

Norwich, John Julius, *A History of Venice*, London 1983.

Nuovo, Isabella, "Galateo, Venezia e Sannazaro", in *La Serenissima e il regno. Nel V centenario dell'Arcadia di Iacopo Sannazaro. Atti del Convegno di studi, Bari-Venezia, 4–8 ottobre 2004*, ed. by Davide Canfora, Angela Caracciolo Aricò, Bari 2006, 533–561.

Nuti, Lucia, "Mapping Places: Chorography and Vision in the Renaissance", in *Mappings*, ed. by Denis Cosgrove, London 1999, 90–108.

Nuti, Lucia, "La pianta di Pisa di Giuliano da Sangallo, 1512–1513", in *Rappresentare la città: topografie urbane nell'Italia di antico regime*, ed. by Marco Folin, Reggio Emilia 2010, 145–156.

Nuti, Lucia "Surveying, Interpreting, and Designing: The Multiple Essence of a Sixteenth-Century Drawing", *Journal of the Society of Architectural Historians*, 75 (2016), 5–24.

Nuvoloni, Laura, Stephen Parkin, Paolo Sachet, "Aldo Manuzio e l'oggetto libro", in *Aldo Manuzio: il rinascimento di Venezia*, catalogue of the exhibition (Venezia, Gallerie dell'Accademia, 19 marzo–19 giugno 2016) ed. by Guido Beltramini, Davide Gasparotto, Venice 2016, 77–90.

Ohlig, Christoph P. J., *De aquis pompeiorum: das Castellum Aquae in Pompeji: Herkunft, Zuleitung und Verteilung des Wassers*, Nijmegen 2001.

Olivieri, Achille, *Erasmo, Venezia e la cultura padana nel '500*, ed. by Achille Olivieri, Rovigo 1995.

Oratores Graeci, ΙΣΟΚΡΑΤΟΥΣ ΛΟΓΟΙ. ΑΛΚΙΔΑΜΑΝΤΟΣ, ΚΑΤΑ ΣΟΦΙΣΤΩΝ. ΓΟΡΓΙΟΥ, ΕΛΕΝΗΣ ΕΓΚΩΜΙΟΝ. ΑΡΙΣΤΕΙΔΟΥ ΠΑΝΑΘΗΝΑΙΚΟΣ, ΤΟΥ ΑΥΤΟΥ ΡΩΜΗΣ ΕΓΚΩΜΙΟΝ. Isocratis Orationes. Alcidamantis contra dicendi magistros. Gorgiae de laudibus Helenae. Aristidis de laudibus Athenarum. Eiusdem de laudibus vrbis Romae, Venice: Aldo Manuzio and Andrea Torresano, *3 May 1513*.

Pagliara, Pier Nicola, "La Roma antica di Fabio Calvo: note sulla cultura antiquaria e architettonica", *Psicon*, 3 (1976), 65–87.

Pagliara, Pier Nicola, "Vitruvio da testo a canone", in *Memoria dell'antico nell'arte italiana*, ed. by Salvatore Settis, vol. III, *Dalla tradizione all'archeologia*, Turin 1984, 2–38.

Pagliara, Pier Nicola, "Giovanni Giocondo", *Dizionario Biografico degli Italiani*, vol. 56, Rome 2001, 326–338.

Pagliara, Pier Nicola, "Fra Giocondo e l'edizione del 'De Architectura' del 1511", in *Giovanni Giocondo umanista, architetto e antiquario*, ed. by Pierre Gros, Pier Nicola Pagliara, Venice 2014, 21–52.

Pane, Roberto, *Il Rinascimento in Italia meridionale*, Milan 1975–1977.

Panofsky, Erwin, "Erasmus and the Visual Arts", *Journal of the Warburg and Courtauld Institutes*, 32 (1969), 200–227.

Panormita, Antonio, *De dictis et factis Alphonsi Regis Aragonum Libri Quator. Commentarium in eosdem Aeneae Sylvj*, Basel: Officina Heruagiana, 1538.

Parma, Aniello, "Universus numerus curiae Pollio Iulio Clementiano statuam conlocavit", *Quaderni Lupiensi di Storia e Diritto*, 5 (2015), 95–107.

Pastoreau, Michel, *Le cochon. Histoire d'un cousin mal aimé*, Paris 2009.

Pedio, Tommaso, *Storia della storiografia del Regno di Napoli*, Chiaravalle 1973.

Pellegrini, Paolo, "Studiare Svetonio a Padova alla fine del Quattrocento", *Incontri triestini di filologia classica*, 7 (2007–2008), 53–64.

Pellegrini, Paolo, "Musuro, Marco", in *Dizionario Biografico degli Italiani*, vol. 77, Rome 2012, 576–582.

Percopo, Erasmo, "Una statua di Tommaso Malvico ed alcuni sonetti del Tebaldeo", *Napoli nobilissima*, 2 (1893), 10–13.

Perini, Giovanna, "Carmi inediti su Raffaello e sull'arte della prima metà del Cinquecento a Roma e Ferrara e il mondo dei Coryciana", *Römisches Jahrbuch der Bibliotheca Hertziana*, 32 (1997–1998), 367–407.

Perotti, Niccolò, *Cornu Copiae, seu linguae Latinae commentarii*, edited by Jean-Luis Charlet *et alii*, 8 vols., Sassoferrato 1989–2001.

Petraglione, Giuseppe, "Pier Candido Decembrio, *De laudibus Mediolanesium urbis panegyricus*", *Archivio Storico Lombardo*, 34 (1907), 27–45.

Pidatella, Chiara, "Cupido di marmo, Cupido di bronzo: nota intorno al materiale dell'Erote antico posseduto da Isabella d'Este", *Acme*, 59 (2006), 243–250.

Pidatella, Chiara, "Loreto, 30 aprile 1512: testamento inedito di Gian Cristoforo Romano", in *Scritti di Historia Nostra per Floriano Grimaldi*, ed. by Maurizio Landolfi, Recanati 2011, 85–99.

Pidatella, Chiara, "Fra Roma, Urbino e Loreto: l'amicizia fra Gian Cristoforo Romano, il Bibbiena e Pietro Bembo (1505–1520)", in *Attraverso la cultura figurativa italiana del Cinquecento: le stagioni del Bembo*, ed. by Vittoria Romani, Marsel Grosso, forthcoming.

Pignatti, Franco, "Il ritratto dell'amata nella lirica del Cinquecento", in *Officine del nuovo. Sodalizi fra letterati, artisti ed editori nella cultura italiana fra Riforma e Confroriforma. Atti del simposio internazionale: Utrecht 8–10 novembre 2007*, ed. by Harald Hendrix, Paolo Procaccioli, Manziana (Rome) 2008, 267–307.

Pinto, Carlo, *Degli epigrammi di don Carlo Pinto Her. Sal. fatti nella morte di don Ferdinando di Castro conte di Lemos viceré di Napoli*, Naples: Giacomo Carlino, 1602.

Pinto, Carlo, *De Vico Garganico apulorum opido Caroli Pinti elegia*, Naples: Giovan Giacomo Carlino, 1607.

Pinto, Carlo, *Dello spaventevole e miserabile avvenimento del fuoco acceso in Monte Vergine nel Regno di Napoli*, Naples: Giovan Giacomo Carlino, 1611.

Polito, Eugenio, *Fulgentibus armis. Introduzione allo studio dei fregi d'armi antichi*, Rome 1998.

Pollard, Graham, *Renaissance Medals. The Collections of the National Gallery of Art Systematic Catalogue*, with the assistance of Eleonora Luciano, Washington D.C. 2007.

Pontano, Giovanni, *De liberalitate, De beneficentia, De magnificentia, De splendore, De conviventia*, Naples: Johann Tresser from Hoestet and Martin from Amsterdam, 1498.

Pontano, Giovanni, *De bello Neapolitano et de Sermone*, ed. by Pietro Summonte, Naples: Sigismund Mayr, May 1509.

Pontano, Giovanni, *Dialoghi*, ed. by Carmelo Privitera, Florence 1943.

Pontano, Giovanni, *I libri delle virtù sociali*, ed. by Tateo, Rome 1999.

Pontano, Giovanni, *Dialogues. Volume 1: Charon and Antonius*, ed. and translated into English by Julia Haig Gaisser, Cambridge (MA)—London 2012.

Ptolemy, *Liber geographiae cum tabulis et universali figura et cum additione locorum quae a recentioribus reperta sunt diligenti cura emendatus et impressus*, ed. by Bernardo Silvano, Venice: Jacobus Pentius de Leucho, 20 March 1511.

Publius Victor, *De Regionibus Urbis Romae libellus aureus*, with a dedicatory elegy by Giano Parrasio, Milan: Giovanni Angelo Scinzenzeler, 1503.

Puppi, Lionello, "Il tempio e gli eroi", in *La grande vetrata di San Giovanni e Paolo: storia, iconologia, restauro*, ed. by Cortà Fumei, Francesco Prosperetti, Serena Romano, Venice 1982, 21–35.

Putnam, Michael C. J., *Jacopo Sannazaro: Latin Poetry*, Cambridge (MA) 2009.

Quasten, John, *Letters of Saint Paulinus of Nola*, ed. by John Quasten, New York 1996.

Quindici, Vincenzo, *Nola antica*, Nola 1984.

Quinterio, Francesco, *Giuliano da Maiano 'grandissimo domestico'*, Rome 1996.

Rabil, Albert Jr., *Knowledge, Goodness, and Power: The Debate over Nobility among Quattrocento Italian Humanists*, Binghamton (NY) 1991.

Ranzano, Pietro, *Descriptio totius Italiae (Annales, XIV–XV)*, ed. by Adele di Lorenzo, Bruno Figliuolo, Paolo Pontari, Rome 2007.

Rauty, Raffaele, *Quando c'erano gli intellettuali. Rileggendo Cultura popolare e marxismo*, ed. by Raffaele Rauty, *Postfazione* by Tullio Seppilli, Milan 2015.

Rea, Giuseppina, *Scavi archeologici e scoperte di antichità nella città di Napoli nella Historia Neapolitana di Fabio Giordano*, PhD thesis, Università degli Studi di Napoli Federico II, Dipartimento di Studi Umanistici, 2011–2012.

Remondini, Gianstefano, *Della Nolana Ecclesiastica Storia*, Naples: Giovanni di Simone, 1747–1757.

Remotti, Francesco, *Morte e trasformazione dei corpi. Interventi di tanatometamòrfosi*, ed. by Francesco Remotti, Milan 2006.

Renouard, Antoine Austin, *Annales de l'imprimerie des Alde, ou histoire des trois Manuce et de leurs éditions*, Paris 1834.

Revest, Clémence, "Les enjeux de la transmission aux origines de l'humanisme. L'exemple de la *Laudatio Florentinae urbis* de Leonardo Bruni", *La transmission, Bulletin de Questes*, 11 (2007), 7–16.

Rijser, David, "The Scultpor as Philologist: Interaction between Scholarship and the Arts in the Goritz Chapel and the 'Coryciana', Rome 1512–1527", in *Officine del nuovo. Sodalizi fra letterati, artisti ed editori nella cultura italiana fra Riforma e Confroriforma. Atti del simposio internazionale: Utrecht 8–10 novembre 2007*, ed. by Harald Hendrix, Paolo Procaccioli, Manziana (Rome) 2008, 257–265.

Robey, David, John Easton Law, "The Venetian Myth and the *Republica Veneta* of Vergeri", *Rinascimento*, 15 (1975), 3–59.

Robotti, Ciro, "Architettura catalana in Carinola", in Mario Rosi, *Carinola: Pompei quattrocentesca*, Naples 1979, 95–118.

Romano, Angelo, "De Ferrariis, Antonio", in *Dizionario Biografico degli Italiani*, vol. 33, Rome 1987, 738–741.

Romano, Serena, "La vetrata dei Santi Giovanni e Paolo: esercizi di attribuzione", *Arte Veneta*, 35 (1981), 41–51.

Romano, Serena, "La vetrata: i maestri e gli artefici", in *La grande vetrata di San Giovanni e Paolo: storia, iconologia, restauro*, ed. by Cortà Fumei, Francesco Prosperetti, Serena Romano, Venice 1982, 51–71.

Romano, Serena, *Ritratto di fanciullo di Girolamo Mocetto*, Modena 1985.

Romano, Serena, "Mocetto", in *The Dictionary of Art*, vol. XXI, London 1996, 748–749.

Romano, Giovanni, Claudio Salsi (eds.), *Maestri della scultura in legno nel ducato degli Sforza*, Catalogue of the exhibition (Milan, 2005–2006), Cinisello Balsamo 2005.

Rosi, Giorgio, "Il Campanile della Cattedrale di Nola", in *Bollettino d'Arte*, s. 4, 34 (1949), 10–20.

Rossi, Francesco, *La Collezione Mario Scaglia. Placchette*, 2 vols., Bergamo 2011.

Roth, Anthony, "The Lombard Sculptor Benedetto Briosco: Works of the 1490s", *The Burlington Magazine*, 122 (1980), 7–22.

Rowland, Ingrid D., "Raphael, Angelo Colocci, and the Genesis of the Architectural Orders", *The Art Bulletin*, 74 (1994), 81–104.

Rowland, Ingrid, *The Culture of the High Renaissance: Ancients and Moderns in Sixteenth-Century Rome*, Cambridge 1998.

Ruggiero, Andrea, *Introduzione*, in Ambrogio Leone, *Nola*, Naples 1997, 7–92.

Russo, Alfonso, *La chiesa dei Santi Apostoli in Nola*, Nola 1973.

Sabba da Castiglione, *Ricordi, overo ammaestramenti [...] ne' quali [...] si ragiona di tutte le materie honorate che si ricercano a un vero gentil'huomo*, Venice: Giovanni Bariletto 1569.

Sabellico, Marco Antonio, *De Venetae urbis situ*, Venice: Damianus de Mediolano, de Gorgonzola, about 1494–1495.

Salatin, Francesca, "La basilica di Fano, Giocondo, Palladio e il Vitruvio Ferrarese", *Annali di Architettura*, 24 (2012), 9–18.

Salatin, Francesca, "'In alcune figure errò Fra Giocondo'. Note su alcune xilografie del Vitruvio del 1511", in *Vitruvio e l'archeologia*, ed. by Paolo Clini, Venice 2014, 175–192.

Sampaolo, Valeria, "Dati archeologici e fenomeni vulcanici nell'area nolana. Nota preliminare", in *Tremblements de terre, éruptions volcaniques et vie des hommes dans la Campanie antique*, ed. by Claude Albore Livadie, Naples 1986, 113–119.

Sampaolo, Valeria, "Nola. Teatro romano", *Bollettino di Archeologia del Ministero per i Beni Culturali ed Ambientali*, 11–12 (1991), 166–167.

Sannazaro, Jacopo, *Arcadia*, ed. by Carlo Vecce, Rome 2013.

Santucci, Giovanni, "Giuliano da Sangallo e il disegno di città. La Pianta di Pisa 7950A della Galleria degli Uffizi", in *Giuliano da Sangallo*, ed. by Amedeo Belluzzi, Caroline Elam, Francesco Paolo Fiore, Milan 2017, 260–275.

Sanudo, Marin, *De origine, situ et magistratibus urbis Venetae, ovvero La città di Venetia (1493–1530)*, ed. by Angela Caracciolo Aricò, Venezia 2011.

Sanudo, Marin, *I Diarii*, ed. by Guglielmo Berchet, Rinaldo Fulin, Federico Stefani, Niccolò Barozzi, Marco Allegri, 58 vol., Venice 1879–1902.

Savino, Eliodoro, *Campania Tardoantica*, Bari 2005.

Scaglia, Gustina, "The Origin of an Archaeological Plan of Rome by Alessandro Strozzi", *Journal of the Warburg and Courtauld Institutes*, 27 (1964), 137–163.

Schipa, Michelangelo, "Alcune opinioni intorno ai seggi o sedili di Napoli nel medioevo", *Napoli Nobilissima*, 15 (1906), 97–99; 113–115.

Schulz, Juergen, "Jacopo de' Barbari's View of Venice: Map Making, City Views, and Moralized Geography before the Year 1500", *The Art Bulletin*, 60 (1978), 425–474.

Scirocco, Elisabetta, *Arredi liturgici dei secoli XI–XIII in Campania: le cattedrali di Salerno, Ravello, Amalfi, Caserta Vecchia, Capua*, PhD Thesis, University Federico II of Naples, 2009.

Servius, *Servii Grammatici qui feruntur in Vergilii carmina commentarii*, ed. by Georg Thilo, 3 vols., Hermann Hagen, Leipzig 1878–1887.

Sica, Francesco, *Ambrogio Leone tra umanesimo e scienze della natura*, Salerno 1983.

van der Sman, Gert Jan, *Le siècle de Titien: gravures vénitiennes de la Renaissance*, Zwolle 2003.

van der Sman, Gert Jan, "Girolamo Mocetto and the Antique", *Print Quarterly*, 30 (2013), 165–170.

Spampanato, Vincenzo, *Vita di Giordano Bruno*, Messina 1921.

Sodano, Rosanna, "Intorno ai 'Coryciana': conflitti politici e letterari in Roma dagli anni di Leone X a quelli di Clemente VII", *Giornale storico della letteratura italiana*, 583 (2001), 420–450.

Sophocles, *ΣΟΦΟΚΛΕΥΣ ΤΡΑΓΩΔΙΑΙ ΕΠΤΑ ΜΕΤ'ΕΞΕΓΗΣΕΩΝ. Sophoclis Tragaediae septem cum commentariis*, Venice: Aldo Manuzio, August 1502.

Spencer, John R., "Ut Rhetorica Pictura: A Study in Quattrocento Theory of Painting", *Journal of the Warburg and Courtald Institutes*, 20 (1957), 26–44.

Speranzi, David, *Marco Musuro. Libri e scrittura*, Rome 2013.

Sponza, Sandro, "La vetrata dei Santi Giovanni e Paolo a Venezia in mostra", *Arte Veneta*, 36 (1982), 289–292.

Sponza, Sandro, *Per una monografia sulla Basilica dei Santi Giovanni e Paolo*, ed. by Sandro Sponza, Venice 1996.

Spruit, Leendert, "Leone, Ambrogio", in *Dizionario Biografico degli Italiani*, vol. 64, Rome 2005, 560 562.

Stefanile, Giacomo, "La 'reggia' Orsini e l'architettura del Rinascimento nell'Italia meridionale", in *Nola e il suo territorio dalla fine del Medio Evo al XVII secolo: momenti di storia culturale e artistica*, ed. by Tobia Toscano, Nola 1996, 45–74.

Summonte, Giovanni Antonio, *Istoria della Città e Regno di Napoli*, books I–II, Naples: Giovan Giacomo Carlino, 1601.

Sutermeister, Guido, *Gli editori 'da Legnano': 1470–1525*, Varese 1946.

Svalduz, Elena, "Tre disegni per una pianta: la 'pianta de Venetia' di Cristoforo Sabbadino (1557)", in *Rappresentare la città. Topografie urbane nell'Italiadi antico regime*, ed. by Marco Folin, Reggio Emilia 2010, 201–224; 420–421.

Syson, Luke, "Reading Faces: Gian Cristoforo Romano's Medal of Isabella d'Este", in *La corte di Mantova nell'età di Andrea Mantegna: 1450–1550*, ed. by Cesare Mozzarelli, Robert Oresko, Leandro Ventura, Rome 1997, 281–294.

Szépe, Helena K., "L'*Hypnerotomachia Poliphili*, l'avventura tra sogno ed erotismo stampata da Aldo Manuzio", in *Aldo Manuzio: il rinascimento di Venezia*, catalogue of the exhibition (Venezia, Gallerie dell'Accademia, 19 marzo–19 giugno 2016) ed. by Guido Beltramini, Davide Gasparotto, Venice 2016, 137–156.

Tafuri, Manfredo, *Venezia e il rinascimento*, Turin 1985.

Tagliaferro, Giorgio, "Mocetto, Girolamo", in *Dizionario Biografico degli Italiani*, vol. 75, Rome 2011, 162–166.

Tansillo, Luigi, *Il Vendemmiatore*, Naples 1534.

Tartuferi, Angelo, *Il maestro del Bigallo e la pittura della prima metà del Duecento agli Uffizi*, Florence 2007.

Tateo, Francesco, "La disputa della nobiltà. Stoicismo e polemica sociale", in Francesco Tateo, *Tradizione e realtà dell'Umanesimo italiano*, Bari 1974, 355–422.

Tateo, Francesco, "Quasi una *comparatio*: Firenze e Venezia in epistole di Antonio Galateo", in *De Florence à Venise: études en l'honneur de Christian Bec*, ed. by François Livi, Carlo Ossola, Paris 2006, 313–325.

Tebaldeo, Antonio, *Sonetti, capituli et egloge*, Milan: Ulderico Scinzenzeler, 1499.

Temanza, Tommaso, *Antica pianta dell'inclita città di Venezia delineata circa la metà del XII secolo, Ed ora per la prima volta pubblicata, ed illustrata. Dissertazione topografico-storico-critica*, Venice: Carlo Palese, 1781.

Toniolo, Federica, "L'arte del libro. Aldo Manuzio e il libro illustrato a Venezia tra Quattrocento e Cinquecento", in *Aldo Manuzio: il rinascimento di Venezia*, catalogue of the exhibition (Venezia, Gallerie dell'Accademia, 19 marzo-19 giugno 2016) ed. by Guido Beltramini, Davide Gasparotto, Venice 2016, 91–106.

Toppi, Nicolò, *Biblioteca Napoletana*, Naples: Antonio Bulifon, 1678.

Tortelli, Giovanni, *Orthographia*, Rome: Ulrich Han and Simon Nicolai Chardella, [after 10 Aug.] 1471a.

Tortelli, Giovanni, *Orthographia*, Venice: Nicolaus Jenson, 1471b.

Tortelli, Giovanni, *Roma antica*, ed. by Luisa Capoduro, Rome 1999.

Toscano, Gennaro, "Sculture del Quattro e Cinquecento a Nola: la committenza Orsini", *Quaderni. Istituto Nazionale di Studi sul Rinascimento Meridionale*, 6 (1989), 117–142.

Toscano, Gennaro, "Il pulpito del Duomo di Nola", *Antologia di belle arti*, 48–51 (1994), 30–37.

Toscano, Gennaro, "La scultura a Nola dagli Orsini agli Albertini", in *Nola e il suo territorio dalla fine del Medio Evo al XVII secolo. Momenti di storia culturale e artistica*, ed. by Tobia R. Toscano, Naples 1996, 85–105.

Toscano, Tobia R., "Il *De Nola* di Ambrogio Leone", in *Nola e il suo territorio: dalla fine del Medio Evo al 17. secolo, momenti di storia culturale e artistica: atti del 2. Corso di formazione per docenti in servizio Didattica e territorio, 10 febbraio–28 maggio 1994*, ed. by Tobia R. Toscano, Nola 1996, 19–23.

Toscano, Tobia R., "Giano Anisio tra Nola e Napoli: amicizie polemiche e dibattiti", in *Nola fuori di Nola. Itinerari italiani ed europei di alcuni nolani illustri*, ed. by Tobia R. Toscano, Castellammare di Stabia 2001, 35–56.

Trout, Dennis E., *Paulinus of Nola: Life, Letters, and Poems*, Berkeley 1999.

Tura, Adolfo, *Fra Giocondo et les textes français de géometrie pratique*, Genève 2008.

Tutini, Camillo, *Dell'origine e fundatione de' seggi di Napoli, del tempo in che furono instituiti, e della separation de' nobili dal popolo*, Naples: Beltrano, 1644.

Valeri, Claudia, "Il tempio augusteo e la cattedrale di San Procolo", in Claudia Valeri, *Marmora Phlegraea: sculture dal Rione Terra*, Rome 2005, 31–42.

Valerio, Vladimiro, *Società, uomini e istituzioni cartografiche nel Mezzogiorno d'Italia*, Florence 1993.

Valerio, Vladimiro, "Astronomia e cartografia nella Napoli aragonese", *Rivista Geografica Italiana*, 100 (1993), 291–303.

Valerio, Vladimiro, "Cartography in the Kingdom of Naples during the Early Modern Period", in *The History of Cartography*, 3.1, *Cartography in the European Renaissance*, ed. by D. Woodward, Chicago 2007, 940–974.

Vasari, Giorgio, *Le vite de' più eccellenti architetti, pittori et scultori italiani, da Cimabue insino a' tempi nostri*, Florence: Lorenzo Torrentino, 1550.

Vasari, Giorgio, *Le vite de' più eccellenti scultori, e architettori*, Florence: Giunti, 1568.

Vecce, Carlo, "Sannazaro e Alberti. Una lettura del *De re aedificatoria*", in *Filologia umanistica: per Gianvito Resta*, ed. by Vincenzo Fera, Giacomo Ferraù, Padua 1997, vol. III, 1821–1860.

Vecce, Carlo, "*Salutate Messer Ambrogio*. Ambrogio Leone entre Venise et l'Europe", *Les Cahiers de l'Humanisme*, 1 (2000), 171–181.

Vecce, Carlo, "Sannazaro lettore del *De re aedificatoria*", in *Alberti e la cultura del Quattrocento, Atti del convegno internazionale del Comitato Nazionale VI centenario della nascita di Leon Battista Alberti (Firenze, 16–18 dicembre 2004)*, ed. by Roberto Cardini, Mariangela Regoliosi, Florence 2007, vol. II, 763–784.

Venturelli, Paola, *Leonardo da Vinci e le arti preziose. Milano tra XV e XVI secolo*, Venice 2002.

Venturi, Adolfo, "Gian Cristoforo Romano", *Archivio Storico dell'Arte*, 1 (1888), 49–59; 107–118; 148–158.

Vincenti, Giovanni, *La contea di Nola dal sec. XIII al XVI. Ricerche storiche e feudali*, Naples 1897.

Vitale, Giuliana, *Araldica e politica. Statuti di ordini cavallereschi 'curiali' nella Napoli Aragonese*, Salerno 1999.

Vitale, Giuliana, *Modelli culturali nobiliari nella Napoli aragonese*, Salerno 2002.

Vitale, Giuliana, *Élite burocratica e famiglia. Dinamiche nobiliari e processi di costruzione statale nella Napoli angioino-aragonese*, Naples 2003.

Vitale, Giuliana, *Ritualità monarchica, cerimonie e pratiche devozionali nella Napoli aragonese*, Salerno 2006.

Vitale, Giuliana, "Sul segretario regio al servizio degli Aragonesi di Napoli", *Studi Storici*, 49 (2008), 293–321.

Vitale, Giuliana, "*Universitates* e *Officiales Regii* in età aragonese nel Regno di Napoli: un rapporto difficile", *Studi Storici*, 51 (2010), 53–72.

Vitale, Giuliana, "A proposito di Amalfitani in Puglia", in *Interscambi socio-culturali ed economici fra le città marinare d'Italia e l'Occidente*, ed. Bruno Figliuolo, Pinuccia F. Simbula, Amalfi 2014, 129–164.

Vitale, Giuliana, *Percorsi urbani nel Mezzogiorno medievale*, Battipaglia 2015.

Vitruvius per Iocundum solito castigatior factus cum figuris et tabula ut iam legi et intelligi possit, [*De architectura*, ed. by Giovanni Giocondo], Venice: Tacuino 1511.

Vitruvius, *De architectura*, ed. by Pierre Gros, Italian translation and notes by Elisa Romano, Antonio Corso, Turin 1997.

Warr, Cordelia, *Dressing for Heaven: Religious Clothing in Italy, 1215–1545*, Manchester 2010.

Weil-Garris Brandt, Kathleen, John F. D'Amico, "The Renaissance Cardinal's Ideal Palace: A Chapter from Cortesi's *De Cardinalatu*", *Memoirs of the American Academy in Rome*, 35 (1980), 45–123.

Weiss, Roberto, "Lineamenti per una storia degli studi antiquari in Italia dal dodicesimo secolo al sacco di Roma del 1527", *Rinascimento*, 9 (1958), 141–201.

Weiss, Roberto, *The Renaissance Discovery of Classical Antiquity*, Oxford 1969.

Weitzmann, Kurt, *Ancient Book Illumination*, Cambridge (MA) 1959.

Welch, Katherine E., *The Roman Amphitheatre. From its Origins to the Colosseum*, Cambridge 2007.

Wilson, Carolyn C., *Renaissance Small Bronze Sculpture and Associated Decorative Arts at the National Gallery of Art*, Washington 1983.

Zambelli, Paola, *Magia bianca, magia nera nel Rinascimento*, Ravenna 2004.

Zambelli, Paola, *White Magic and Black Magic in the European Renaissance*, Leiden-Boston 2007.

Zarmakoupi, Mantha, "Porticus and Cryptoporticus in Roman Luxury Villas: Architectural Design and Cultural Implications", in *Art, Industry and Infrastructure in Roman Pompeii*, ed. by Eric Poehler, Miko Flohr, Kevin Cole, Oxford 2011, 50–181.

Zemon Davis, Natalie, *Society and Culture in Early Modern France: eight essays*, Stanford 1986.

Index of Names

(the name of Ambrogio Leone is not included)

Printed in the United States
By Bookmasters